"RIVETING . . . *UNDER A WHITE SKY* EXPERTLY MIXES TRAVELOGUE, SCIENCE REPORTING AND EXPLANATORY JOURNALISM."

—*THE WASHINGTON POST*

"TO BE A WELL-INFORMED CITIZEN OF PLANET EARTH, YOU NEED TO READ ELIZABETH KOLBERT."

—*ROLLING STONE*

"A SUPERB AND HONEST REFLECTION OF OUR EXTRAORDINARY TIME."

—*NATURE*

"[*UNDER A WHITE SKY*] EXHIBITS KOLBERT'S SCULPTOR-LIKE SKILL FOR MAKING CLIMATE CHANGE FEEL TANGIBLE, HAPPENING BEFORE OUR EYES AND BENEATH OUR FINGERS."

—*WIRED*

"Beautifully and insistently, Kolbert shows us that it is time to think radically about the ways we manage the environment." —Helen Macdonald, *The New York Times Book Review*

Praise for *Under a White Sky*

"A fascinating survey of novel attempts to manage natural systems of all sizes, from preserving tiny populations of desert fish to altering the entire atmosphere . . . One of the great science journalists, Kolbert has for many years been an essential voice, a reporter from the front lines of the environmental crisis. Her new book crackles with the realities of living in an era that has sounded the death knell for our commonly held belief that one can meaningfully distinguish between nature and humanity. . . . [Kolbert] has a marvelous eye for the quirky . . . and she wields figurative language in truly glorious ways. . . . Time to work with what we have, using the knowledge we have, with our eyes fully open to the realities of where we are."

—Helen Macdonald, *The New York Times Book Review*

"Kolbert reveals the Anthropocene at its most absurd. . . . *Under a White Sky* expertly mixes travelogue, science reporting and explanatory journalism, all with the authority of a writer confident enough to acknowledge ambiguity."

—Carlos Lozada, *The Washington Post*

"[*Under a White Sky*] is a tribute to Kolbert's skills as a storyteller that she transforms the quest to deal with the climate crisis into a darkly comic tale of human hubris and imagination that could either end in flames or in a new vision of Paradise."

—Jeff Goodell, *Rolling Stone*

"What makes *Under a White Sky* so valuable and such a compelling read is Kolbert tells by showing. Without beating the reader over the head, she makes it clear how far we already are from a world of undisturbed, perfectly balanced nature—and how far we must still go to find a new balance for the planet's future that still has us humans in it."

—NPR

"From the Mojave to lava fields in Iceland, Kolbert takes readers on a globe-spanning journey to explore these projects while weighing their pros, cons, and ethical implications."

—*The Nation*

"Elizabeth Kolbert's beat is examining the impact of humans on the environment and she does it better than basically everyone. [Kolbert] takes this damage as a starting point rather than a focus, asking how to reconstruct, preserve, and even save nature, going forward."

—*Literary Hub*

"An eye-opening—and at times terrifying—examination of just how far scientists have already gone in their attempts to re-engineer the planet."

—Amy Brady, *Gizmodo*

"This intimate natural history is both a sober assessment of the ecosystems we have harmed and an exciting description of some of the discoveries that could help undo that damage."

—*Scientific American*

"Beautifully written . . . Elizabeth Kolbert is a top journalist."

—Ken Caldeira, *Science*

"Kolbert has a keen awareness of unintended consequences, and she's funny. If you like your apocalit with a side of humor, she will have you laughing while Rome burns."

—Leah C. Stokes, *MIT Technology Review*

"Kolbert's prose is peppered with . . . mordant observations, which bring out the humanity (or animality) in her subjects."

—Ben Cooke, *The Times* (U.K.)

"A meticulously researched and deftly crafted work of journalism that explores some of the biggest challenges of our age."

—Jonathan Watts, *The Guardian*

"Kolbert covers interventions on the cutting edge of science, such as 'assisted evolution,' which would help coral reefs endure warmer oceans. Her style of immersive journalism (which involves being hit by a jumping carp, observing coral sex, and watching as millennia-old ice is pulled from the ice sheets of Greenland) makes apparent the challenges of 'the whole-earth transformation' currently underway. This investigation of global change is brilliantly executed and urgently necessary."

—*Publishers Weekly* (starred review)

"A master elucidator, Kolbert is gratifyingly direct as she assesses our predicament between a rock and a hard place, creating a clarion and invaluable 'book about people trying to solve problems created by people trying to solve problems.'"

—*Booklist* (starred review)

"Urgent, absolutely necessary reading as a portrait of our devastated planet . . . Every paragraph of Kolbert's books has a mountain of reading and reporting behind it."

—*Kirkus Reviews* (starred review)

"A tale not of magic-bullet remedies where maybe this time things will be different when we intervene in nature, but rather of deploying a panoply of strategies big and small in hopes that there is still time to make a difference and atone for our past. A sobering and realistic look at humankind's perhaps misplaced faith that technology can work with nature to produce a more livable planet."

—*Library Journal* (starred review)

By Elizabeth Kolbert

The Sixth Extinction

Field Notes from a Catastrophe

The Prophet of Love

Under a White Sky

UNDER A WHITE SKY

The Nature of the Future

ELIZABETH KOLBERT

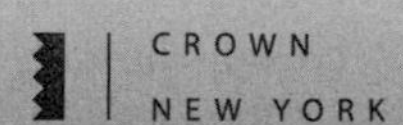

2022 Crown Trade Paperback Edition

Published in the United States by Crown, an imprint of Random House, a division of Penguin Random House LLC, New York.

Crown and the Crown colophon are registered trademarks of Penguin Random House LLC.

Originally published in hardcover in the United States by Crown, an imprint of Random House, a division of Penguin Random House LLC, New York, in 2021.

Portions of this work originally appeared in *The New Yorker*.

Library of Congress Cataloging-in-Publication Data

Names: Kolbert, Elizabeth, author.
Title: Under a white sky / Elizabeth Kolbert.
Description: First edition. | New York: Crown, [2021] | Includes bibliographical references and index.
Identifiers: LCCN 2020047398 (print) | LCCN 2020047399 (ebook) | ISBN 9780593136287 | ISBN 9780593238776 | ISBN 9780593136294 (ebook)
Subjects: LCSH: Nature—Effect of human beings on. | Human ecology. | Environmental protection. | Ecological engineering. | Sustainability.
Classification: LCC GF75 .K65 2021 (print) | LCC GF75 (ebook) | DDC 304.2/8—dc23
LC record available at lccn.loc.gov/2020047398
LC ebook record available at lccn.loc.gov/2020047399

Printed in the United States of America on acid-free paper

crownpublishing.com

456789

Book design by Simon M. Sullivan

To my boys

Sometimes he runs his hammer along the walls, as though to give the signal to the great waiting machinery of rescue to swing into operation. It will not happen exactly in this way—the rescue will begin in its own time, irrespective of the hammer—but it remains something, something palpable and graspable, a token, something one can kiss, as one cannot kiss rescue.

Franz Kafka

Contents

Under a White Sky

DOWN THE RIVER

1

Rivers make good metaphors—too good, perhaps. They can be murky and charged with hidden meaning, like the Mississippi, which to Twain represented "the grimmest and most dead-earnest of reading matter." Alternatively, they can be bright and clear and mirror-like. Thoreau set off for a week on the Concord and Merrimack Rivers and within a day found himself lost in reflection over the reflections he saw playing on the water. Rivers can signify destiny, or coming into knowledge, or coming upon that which one would rather not know. "Going up that river was like traveling back to the earliest beginnings of the world, when vegetation rioted on the earth," Conrad's Marlow recalls. They can stand for time, for change, and for life itself. "You can't step into the same river twice," Heraclitus is supposed to have said, to which one of his followers, Cratylus, is supposed to have replied, "You can't step into the *same* river even once."

It is a bright morning following several days of rain, and the not-quite-river I am riding is the Chicago Sanitary and Ship Canal. The canal is a hundred and sixty feet wide and runs as straight as a ruler. Its waters, the shade of old cardboard, are flecked with candy wrappers and bits of Styrofoam. On this particular morning, traffic consists of barges hauling sand, gravel, and petrochemicals. The one exception is the vessel I'm on, a pleasure craft named *City Living*.

City Living is outfitted with off-white banquettes and a canvas awning that snaps smartly in the breeze. Also on board are the boat's captain and owner and several members of a group called Friends of the Chicago River. The Friends are not a fastidious bunch. Often their outings involve wading knee-deep in polluted water to test for fecal coliform. Still, our expedition is slated to take us farther down the canal than any of them has ever been before. Everyone is excited and, if truth be told, also a little creeped out.

We have made our way into the canal from Lake Michigan, via the Chicago River's South Branch, and now are motoring west, past mountains of road salt, mesas of scrap metal, moraines of rusted shipping containers. Just beyond the city limits, we skirt the outflow pipes of the Stickney plant, said to be the largest sewage operation in the world. From the deck of *City Living*, we can't see the Stickney, but we can smell it. Conversation turns to the recent rains. These have overwhelmed the region's water-treatment system, resulting in "combined sewer overflows," or CSOs. There is speculation about what sort of "floatables" the CSOs have set adrift. Someone wonders if we'll encounter any Chicago River whitefish, local slang for used condoms. We chug on. Eventually, the Sanitary and Ship Canal joins up with another canal, known as the Cal-Sag. At the meeting of the waters, there's a V-shaped park, featuring picturesque waterfalls. Like just about everything else on our route, the waterfalls are manufactured.

If Chicago is the City of the Big Shoulders, the Sanitary and Ship Canal might be thought of as its Oversized Sphincter. Before it was dug, all of the city's waste—the human excrement, the cow manure, the sheep dung, the rotting viscera from the stockyards—ran into the Chicago River, which, in some spots, was so thick with filth it was said a chicken could walk from one bank to the other without getting her feet wet. From the river, the muck flowed into Lake Michigan. The lake was—and remains—the city's sole source of drinking water. Typhoid and cholera outbreaks were routine.

The canal, which was planned in the closing years of the nineteenth century and opened at the start of the twentieth, flipped the river on its head. It compelled the Chicago to change its direction, so that instead of draining into Lake Michigan, the city's ordure would flow away from it, into the Des Plaines River, and from there into the Illinois, the Mississippi, and, ultimately, the Gulf of Mexico. WATER IN CHICAGO RIVER NOW RESEMBLES LIQUID, ran the headline in *The New York Times*.

The reversal of the Chicago was the biggest public-works project of its time, a textbook example of what used to be called, without irony, the control of nature. Excavating the canal took seven years and entailed the invention of a whole new suite of technologies—the Mason & Hoover Conveyor, the Heidenreich Incline—which, together, became known as the Chicago School of Earth Moving. In total, forty-three million cubic yards of rock and soil were gouged out, enough, one admiring commentator calculated, to build an island more than fifty feet high and a mile square. The river made the city, and the city remade the river.

But reversing the Chicago didn't just flush waste toward St. Louis. It also upended the hydrology of roughly two-thirds of the United States. This had ecological consequences, which had financial consequences, which, in turn, forced a whole new round of interventions on the backward-flowing river. It is toward these that *City Living* is cruising. We're approaching cautiously, though maybe not cautiously enough, because at one point *City Living* almost gets squished between two double-wide barges. The deckhands yell down instructions that are initially incomprehensible, then become unprintable.

About thirty miles up the down river—or is it down the up river?—we draw near our goal. The first sign that we're getting close is a sign. It's the size of a billboard and the color of a plastic lemon. WARNING, it says. NO SWIMMING, DIVING, FISHING, OR MOORING. Almost immediately there's another sign, in white: SUPERVISE ALL PASSENGERS, CHILDREN, AND PETS. Several hundred yards farther along, a third sign appears, maraschino red. DANGER, it states. ENTERING ELECTRIC FISH BARRIERS. HIGH RISK OF ELECTRIC SHOCK.

Everyone pulls out a cell phone or a camera. We photograph the water, the warning signs, and each other. There's joking on board that one of us should dive into the river electric, or at least stick a hand in to see what happens. Six great blue herons, hoping for an easy dinner, have gathered, wing to wing, on the bank, like students waiting on line in a cafeteria. We photograph them, too.

That man should have dominion "over all the earth, and over every creeping thing that creepeth upon the earth," is a prophecy that has hardened into fact. Choose just about any metric you want and it tells the same story. People have, by now, directly transformed more than half the ice-free land on earth—some twenty-seven million square miles—and indirectly half of what remains. We have dammed or diverted most of the world's major rivers. Our fertilizer plants and legume crops fix more nitrogen than all terrestrial ecosystems combined, and our planes, cars, and power stations emit about a hundred times more carbon dioxide than volcanoes do. We now routinely cause earthquakes. (A particularly damaging human-induced quake that shook Pawnee, Oklahoma, on the morning of September 3, 2016, was felt all the way in Des Moines.) In terms of sheer biomass, the numbers are stark-staring: today people outweigh wild mammals by a ratio of more than eight to one. Add in the weight of our domesticated animals—mostly cows and pigs—and that ratio climbs to twenty-two to one. "In fact," as a recent paper in the *Proceedings of the National Academy of Sciences* observed, "humans and livestock outweigh all vertebrates combined, with the exception of fish." We have become the major driver of extinction and also, probably, of speciation. So pervasive is man's impact, it is said that we live in a new geological epoch—the Anthropocene. In the age of man, there is nowhere to go, and this includes the deepest trenches of the oceans and the middle of the Antarctic ice sheet, that does not already bear our Friday-like footprints.

An obvious lesson to draw from this turn of events is: be careful what you wish for. Atmospheric warming, ocean warming, ocean acidification, sea-level rise, deglaciation, desertification, eutrophication—these are just some of the by-products of our species's success. Such is the pace of what is blandly labeled "global change" that there are only a handful of comparable examples in earth's history, the most recent being the asteroid impact that ended the reign of the dinosaurs, sixty-six million years ago. Humans are producing no-analog climates, no-analog ecosystems, a whole no-analog future. At this point it might be prudent to scale back our commitments and reduce our impacts. But there are so many of us—as of this writing nearly eight billion—and we are stepped in so far, return seems impracticable.

And so we face a no-analog predicament. If there is to be an answer to the problem of control, it's going to be more control. Only now what's got to be managed is not a nature that exists—or is imagined to exist—apart from the human. Instead, the new effort begins with a planet remade and spirals back on itself—not so much the control of nature as the *control of* the control of nature. First you reverse a river. Then you electrify it.

The United States Army Corps of Engineers has its Chicago District headquarters in a Classical Revival building on LaSalle Street. A plaque outside the building explains that it was the site of the General Time Convention of 1883, held to sync the country's clocks. The process involved pruning dozens of regional time zones down to four, which, in many towns, resulted in what's become known as the "day with two noons."

Since its founding, under President Thomas Jefferson, the Corps has been dedicated to out-scaled interventions. Among the many world-altering undertakings it's had a shovel in are: the Panama Canal, the St. Lawrence Seaway, the Bonneville Dam, and the Manhattan Project. (To build the atomic bomb, the Corps created a new division; it called this the Manhattan District to disguise the project's true purpose.) It is a sign of the times that the Corps finds itself increasingly involved in backward-looping, second-order efforts, like managing the electric barriers on the Sanitary and Ship Canal.

One morning not long after my boat trip with the Friends, I visited the Corps' Chicago office to talk with the engineer in charge of the barriers, Chuck Shea. The first thing I noticed on arriving was a pair of giant Asian carp, mounted on rocks, next to the reception desk. Like all Asian carp, they had eyes near the bottom of their heads, so it looked as if they'd been mounted upside down. In a curious commingling of fake fauna, the plastic fish were surrounded by little plastic butterflies.

"I never would have pictured when I was studying engineering years ago that I would spend so much time thinking about a fish," Shea told me. "But, actually, it's pretty good for party conversation." Shea is a slight man with graying hair, wire-rimmed glasses, and the diffidence that comes from dealing with problems words can't solve. I asked him how the barriers worked, and he stuck out his hand, as if to shake mine.

"We pulse electricity into the waterway," he explained. "And basically you just have to transmit enough electricity to the water to ensure that you're getting an electrical field throughout the area.

"The electric-field strength is increasing as you move from upstream to downstream or vice versa, so if my hand were a fish, its nose is here," he continued, indicating the tip of his middle finger, "and its tail is here." He pointed to the base of his palm, then set the outstretched hand wiggling.

"What happens is, the fish is swimming in, and its nose is experiencing one electrical voltage, and its tail is experiencing another. That's what makes the current actually flow through the body. It's the current flowing through a fish that will shock them or electrocute them. So a big fish has a big voltage difference from its nose to its tail. A smaller fish doesn't have that much distance for the voltage to cover, so the shock is smaller."

He sat back and dropped his hand into his lap. "The good news is that Asian carp are very big fish. They're public enemy number one." A person, I noted, is pretty big, too. "All people react differently to electricity," Shea replied. "But the bottom line, unfortunately, is that it can be fatal."

Shea told me the Corps had gotten into the barrier business in the late 1990s, thanks to a push from Congress. "It was a fairly open-ended directive," he said. " 'Do something!' "

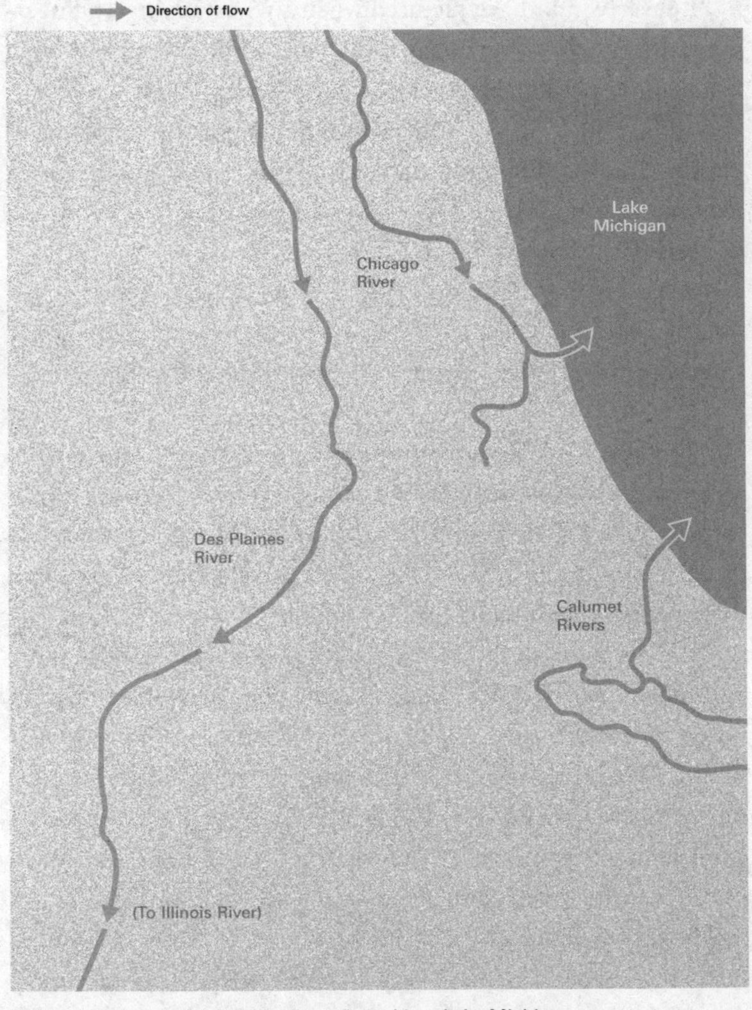

Before its reversal, the Chicago River flowed into Lake Michigan.

The task set for the Corps was a tricky one: to make the Sanitary and Ship Canal impassable for fish, without impeding the movement of people, their cargo, or their waste. The Corps considered more than a dozen possible approaches, including: dosing the canal with poison, irradiating it with ultraviolet light,

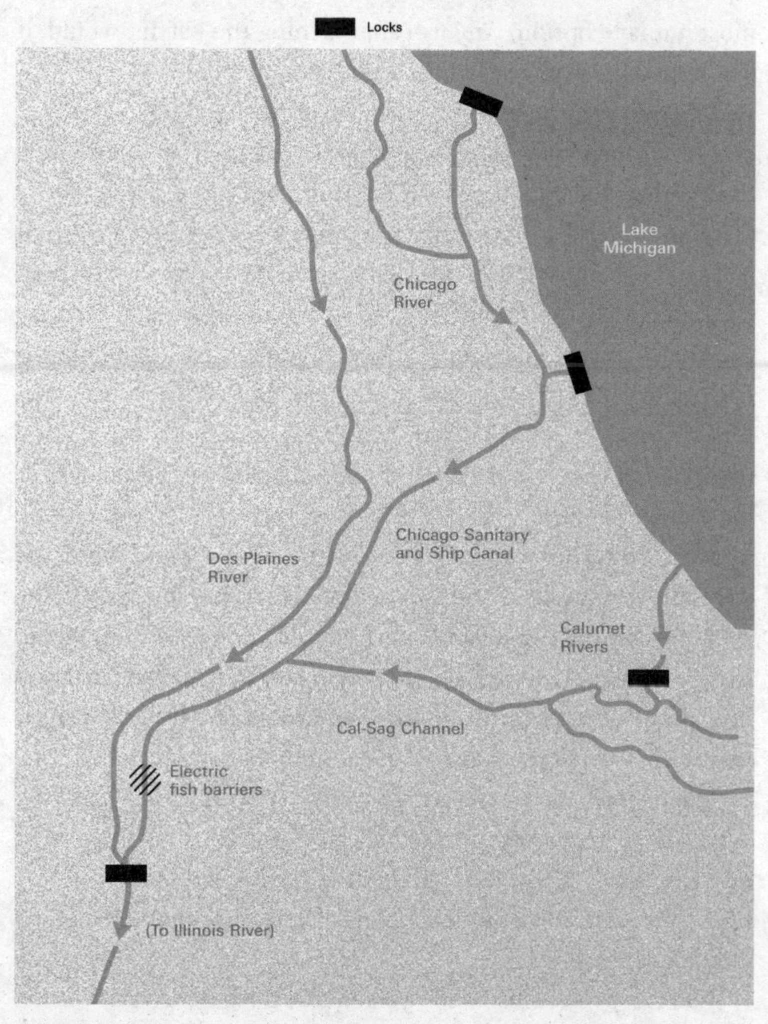

The Chicago Sanitary and Ship Canal redirected the river away from the lake.

zapping it with ozone, using power-plant effluent to heat the water, and installing giant filters. It even looked into loading the canal with nitrogen to create the sort of anoxic environment typically associated with raw sewage. (This last option was rejected in part owing to its cost—an estimated $250,000 a day.) Electrification won out because it was cheap and seemed the most humane option. Any fish approaching the barrier would, it was hoped, be repelled before it was actually killed.

The first electric barrier went live on April 9, 2002. The species it was originally supposed to repel was a frog-faced interloper called the round goby. The round goby is a native of the Caspian Sea and an aggressive consumer of other fishes' eggs. It had established itself in Lake Michigan, and the fear was it would use the Sanitary and Ship Canal to swim out of the lake and into the Des Plaines River. From there, it could swim into the Illinois River and on to the Mississippi. But, as Shea put it to me, "Before the project could be activated, the round goby was already on the other side." It became a case of electrifying the canal after the fish had bolted.

Meanwhile, other invaders—Asian carp—were moving in the opposite direction, up the Mississippi, toward Chicago. If the carp got through the canal, they would, it was feared, wreak havoc in Lake Michigan, before moving on to wreak more havoc in Lakes Superior, Huron, Erie, and Ontario. One Michigan politician warned the fish could "ruin our way of life."

"Asian carp are a very good invasive species," Shea told me. Then he corrected himself: "Well, not 'good'—they're good at being invasive. They're adaptable and they're able to thrive in a lot of different environments. And that's what makes them so difficult to deal with."

The Corps later installed two additional barriers on the canal, which significantly upped the voltage, and, at the time of my visit, it was replacing the original barrier with a more powerful version. It was also planning to take the fight to a whole new level, by installing a barrier that featured loud noise and bubbles. The cost of the bubble barrier was first estimated at $275 million, then later rose to $775 million.

"People joke about it being a disco barrier," Shea said. It was a line, it occurred to me, he might well have used at a party.

Though people often talk about Asian carp as if it were a single species, the term is a catchall for four fish. All four are native to China, where they're referred to collectively as 四大家鱼, a phrase that translates into English roughly as the "four famous domestic fishes." The Chinese raise the famous four together in ponds and have been doing so since the thirteenth century. The practice has been called "the first documented example of integrated polyculture in human history."

Each of the famous four has its own special talent, and when they join forces, they are, like the Fantastic Four, pretty much unstoppable. Grass carp (*Ctenopharyngodon idella*) eat aquatic plants. Silver carp (*Hypophthalmichthys molitrix*) and bighead carp (*Hypophthalmichthys nobilis*) are filter feeders; the two fish suck water in through their mouths and then rake out the plankton using comb-like structures in their gills. Black carp (*Mylopharyngodon piceus*) eat mollusks, like snails. Throw farm clippings into a pond and the grass carp will eat them. Their waste will promote algae growth. The algae will then feed silver carp and also tiny aquatic animals, like water fleas, the preferred diet of bighead carp. This system has allowed the Chinese to harvest immense quantities of carp—almost fifty billion pounds in 2015 alone.

In the sort of irony the Anthropocene teems with, the number of free-swimming carp in China has crashed even as pond-raised populations have soared. Thanks to projects like the Three Gorges Dam, on the Yangtze, river fish are having trouble spawning. The carp are thus at once instruments of human control and victims of it.

The four famous fish ended up in the Mississippi, at least in part, owing to *Silent Spring*—another Anthropocene irony. In the book, whose working title was *The Control of Nature*, Rachel Carson denounced the very idea.

"The 'control of nature' is a phrase conceived in arrogance, born of the Neanderthal age of biology and philosophy, when it was supposed that nature exists for the convenience of man," she wrote. Herbicides and pesticides represented the very worst kind of "cave man" thinking—a club "hurled against the fabric of life."

The indiscriminate application of chemicals was, Carson warned, harming people, killing birds, and turning the country's waterways into "rivers of death." Instead of promoting pesticides and herbicides, government agencies ought to be eliminating them; "a truly extraordinary variety of alternatives" were available. An alternative Carson particularly recommended was setting one biological agent against another. For instance, a parasite could be imported to feed on an unwanted insect.

"In that book the problem—the villain—was the broad, almost unrestricted use of chemicals, particularly the chlorinated hydrocarbons, like DDT," Andrew Mitchell, a biologist at an aquaculture research center in Arkansas who's studied the history of Asian carp in America, told me. "So that's the context of all this: How are we going to get rid of this heavy chemical usage and still have some sort of control? And that probably has as much to do with the importation of carp as anything. These fish were biological controls."

One year after *Silent Spring*'s publication, in 1963, the U.S. Fish and Wildlife Service brought the first documented shipment of Asian carp to America. The idea was to use the carp, much as Carson had recommended, to keep aquatic weeds in check. (Weeds like Eurasian watermilfoil—another introduced species—can clog lakes and ponds so thoroughly that boats or even swimmers can't get through.) The fish were baby grass carp—"fingerlings"—and they were raised at the agency's Fish Farming Experimental Station in Stuttgart, Arkansas. Three years later, biologists at the station succeeded in getting one of the carp—now grown—to spawn. Thousands more fingerlings resulted. Pretty much immediately, some escaped. Baby carp made their way into the White River, a tributary of the Mississippi.

Later, in the 1970s, the Arkansas Game and Fish Commission found a use for silver and bighead carp. The Clean Water Act had just been passed, and local governments were under pressure to comply with the new standards. But a lot of communities couldn't afford to upgrade their sewage-treatment plants. The Game and Fish Commission thought that stocking carp in treatment ponds might help. The carp would reduce the nutrient load in the ponds by consuming the algae that thrived on the excess nitrogen. For one study, silver carp were placed in treatment lagoons in Benton, a suburb of Little Rock. The fish did indeed reduce the nutrient load before they, too, escaped. No one is quite sure how, because no one was watching.

"At the time, everybody was looking for a way to clean up the environment," Mike Freeze, a biologist who worked with carp at the Arkansas Game and Fish Commission, told me. "Rachel Carson had written *Silent Spring*, and everybody was concerned about all the chemicals in the water. They weren't nearly as concerned about non-native species, which is unfortunate."

• • •

The fish—mostly silver carp—lay in a bloody heap. There were scores of them, and they'd been tossed alive into the boat. I'd been watching them pile up for hours, and while the ones at the bottom were, I figured, by now dead, those on top continued to gasp and thrash. I thought I could detect an accusatory glint in their low-set eyes, but I had no idea if they could even see me or whether this was just projection.

It was a sultry summer morning a few weeks after my trip on *City Living*. The gasping carp, a trio of biologists employed by the state of Illinois, several fishermen, and I were all bobbing on a lake in the town of Morris, about sixty miles southwest of Chicago. The lake had no name, having started off as a gravel pit. To get access to it, I'd had to sign a release form from the company that owned it, stating that, among other things, I was not carrying any firearms and would not smoke or use "flame-producing devices." The form showed the outline of the pit-turned-lake, which looked like a child's drawing of a tyrannosaurus. Where the tyrannosaurus's navel would be, if tyrannosauruses had had navels, was a channel linking the lake to the Illinois River. This arrangement accounted for the carp. Carp need moving water to reproduce—either that or injections of hormones—but once they're done spawning, they like to retreat to slack water to feed.

Morris might be thought of as the Gettysburg in the war against Asian carp. South of the town, the carp are legion; north of it they are rare (though how rare is a matter of debate). A great deal of time, money, and fish flesh are devoted to trying to keep things this way. The process is known as "barrier defense," and it's supposed to prevent large carp from reaching the electric barriers. If electrocution were a fail-safe deterrent, then barrier defense wouldn't be necessary, but no one I spoke to, and this included officials like Shea, at the Army Corps of Engineers, seemed eager to see the technology put to the test.

"Our goal is to keep carp out of the Great Lakes," one of the biologists told me as we floated over the former gravel pit. "We're not depending on the electric barriers."

At the start of the day, the fishermen had set out hundreds of yards of gill net, which they were now pulling in from three aluminum boats. Native fish—flathead catfish, say, or freshwater drum—that got caught in the net were disentangled and tossed back into the lake. Asian carp got thrown into the center of the boats to die.

In the nameless lake, the supply of carp seemed endless. My clothes and also my notebook and tape recorder were spattered with blood and slime. No sooner were the nets hauled in than they were reset. When the fishermen needed to get from one end of their boats to the other, they'd simply wade through the writhing carp in the middle. "Who hears the fishes when they cry?" Thoreau asked. "It will not be forgotten by some memory that we were contemporaries."

The very same qualities that have made the "domestic fishes" famous in China have made them infamous in the United States. A well-fed grass carp can weigh more than eighty pounds. In a single day it can eat almost half of its body weight, and it lays hundreds of thousands of eggs at a time. Bigheads can, on occasion, weigh as much as a hundred pounds. They have bulging brows and look as if they were nursing a grudge. Lacking a true stomach, they feed more or less continuously.

Silver carp are equally voracious; they're such effective filter feeders that they can strain out plankton down to four microns across—a quarter of the width of the finest human hair. Just about wherever they show up, the carp outcompete the native fish until they're practically all that's left. As the journalist Dan

Egan has put it: "Bighead and silver carp don't just invade ecosystems. They conquer them." On the Illinois River, Asian carp currently make up almost three-quarters of the fish biomass, and on some waterways the proportion is even higher. The ecological damage, meanwhile, extends beyond fish; black carp, which feed on mollusks, are, it's feared, pushing already-threatened freshwater mussels over the edge.

"North America has the most diverse assemblage of mussels of any place in the world," Duane Chapman, a research biologist with the U.S. Geological Survey who specializes in Asian carp, told me. "Many species are endangered or already extinct. And now we've essentially dumped the world's most efficient freshwater molluscivore on some of the most endangered mollusks."

One of the fishermen I met in Morris, Tracy Seidemann, was wearing a pair of waterproof overalls that were smeared with gore and a T-shirt with the sleeves cut off. I noticed that he had a tattoo of a carp on one of his sunburned arms. It was, Seidemann told me, a common carp. Common carp, too, are invasive. They were introduced from Europe back in the 1880s and probably wreaked their own kind of havoc. But they've been around so long, people have grown accustomed to them. "I should have put an Asian carp there, I guess," he said, shrugging.

Seidemann told me he used to catch mainly buffalo, which are native to the Mississippi River and its tributaries. (Buffalo look a bit like carp but belong to an entirely different family.) When Asian carp arrived, buffalo populations plummeted. Now Seidemann makes most of his income from contract killing for the Illinois Department of Natural Resources. It seemed rude to ask him how much, but later I learned that contract fishermen can gross more than $5,000 a week.

At the end of the day, Seidemann and the others loaded their boats on trailers and, with the carp still in them, drove into town. The fish, now inert and glassy-eyed, were dumped into a waiting semi-trailer.

This round of barrier defense continued for another three days. The final tally was six thousand four hundred and four silver carp and five hundred and forty-seven bighead. Collectively, the fish weighed more than fifty thousand pounds. They were shipped west in the semi, to be ground into fertilizer.

The Mississippi River's drainage basin is the third largest in the world, exceeded in area only by the Amazon's and the Congo's. It stretches over more than 1.2 million square miles and encompasses thirty-one states and slices of two Canadian provinces. The basin is shaped a bit like a funnel, with its spout sticking into the Gulf of Mexico.

The Great Lakes' drainage basin is also vast. It extends over three hundred thousand square miles and contains eighty percent of North America's fresh surface water supply. This system, which has the shape of an overfed seahorse, drains east into the Atlantic, by way of the St. Lawrence River.

The two great basins abut each other, but they are—or were—distinct aquatic worlds. There was no way for a fish (or a mollusk or a crustacean) to climb out of one drainage system and into the other. When Chicago solved its sewage problem by digging the Sanitary and Ship Canal, a portal opened up, and the two aquatic realms were connected. For most of the twentieth century, this wasn't much of an issue; the canal, loaded with Chicago's waste, was too toxic to serve as a viable route. With the passage of the Clean Water Act and the work of groups like the Friends of the Chicago River, conditions improved, and creatures like the round goby began to slip through.

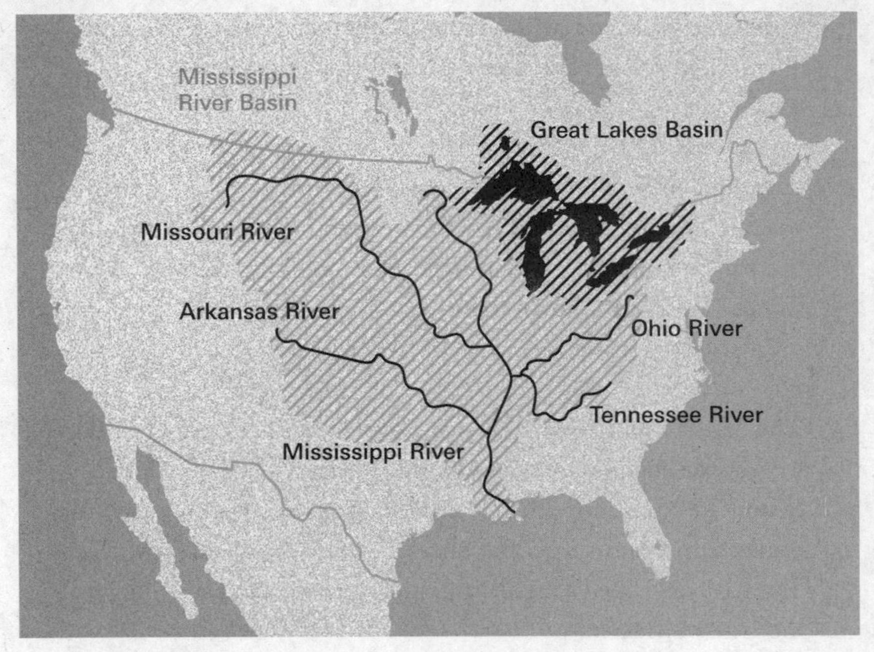

The Chicago River's reversal connected two great drainage basins.

In December 2009, the Corps shut down one of the electrical barriers on the canal to perform routine maintenance. The nearest Asian carp was believed to be fifteen miles downstream. Still, as a precaution, the Illinois Department of Natural Resources dosed the water with two thousand gallons of poison. The result was fifty-four thousand pounds of dead fish. In the mix, one Asian carp—a twenty-two-inch-long bighead—was discovered. Doubtless many fish had sunk to the bottom before they could be netted. Were there more Asian carp among them?

The reaction from neighboring states was fierce. Fifty members of Congress signed a letter to the Corps, expressing their dismay. "There may be no greater threat to the ecosystem of the Great Lakes than the introduction of the Asian carp," the letter said. Michigan filed a lawsuit, demanding that the link between the drainage systems be broken. The Corps studied the options and then, in 2014, released a two-hundred-thirty-two-page report.

According to the Corps' assessment, reimposing "hydrologic separation" would, indeed, be the most effective way to keep carp out of the Great Lakes. It would also, in the Corps' estimate, take twenty-five years—three times as long as the original digging of the canal had—and cost up to $18 billion.

Many experts I spoke to said the billions would be money well spent. They pointed out that each of the two drainage basins has its own roster of invasives, some, like the carp, brought over intentionally, but most introduced accidentally, in ballast water. On the Mississippi side, these include: Nile tilapia, Peruvian watergrass, and convict cichlid from Central America. On the Great Lakes side are: sea lamprey, threespine stickleback, fourspine stickleback, spiny waterflea, fishhook waterflea, New Zealand mud snail, European valve snail, European ear snail, greater European pea clam, humpbacked pea clam, Henslow pea clam, red swamp crayfish, and bloody red shrimp. The surest way to control the invaders would be to plug the canal.

But no one who spoke up for "hydrologic separation" said they thought it would ever happen. To re-replumb Chicago would mean rerouting the city's boat traffic, redesigning its flood controls, and revamping its sewage-treatment system. There were too many constituencies with a vested interest in the way things were. "Politically, it just would never move," the leader of one group that had pushed for separation but had eventually given up on the idea told me. It was a lot easier to imagine changing the river once again—with electricity and bubbles and noise and anything else anyone could dream up—than changing the lives of the people around it.

• • •

The first time I got hit by a carp was near the town of Ottawa, Illinois. It felt like someone had slammed me in the shin with a Wiffle-ball bat.

What people really notice about Asian carp—what literally leaps out at them—is that silver carp jump. One noise that sends them jumping is the thrum of an outboard motor, so waterskiing in carp-infested areas of the Midwest has become its own version of an extreme sport. The sight of silver carp arcing through the air is at once beautiful—like attending a piscine ballet—and terrifying—like facing incoming fire. One of the fishermen I met in Ottawa told me he'd been knocked unconscious by an encounter with a flying carp. A second said he'd long ago lost track of his carp-related injuries, because "you pretty much get hit every day." A woman I read about was knocked off her Jet Ski by a carp and survived only because a passing boater noticed her life jacket bobbing in the water. Countless videos of carp acrobatics

Silver carp, when startled, fling themselves out of the water.

are available on YouTube, with titles like "Asian Carpocalypse" and "The Attack of the Jumping Asian Carp." The town of Bath, Illinois, which sits on a particularly carp-rich stretch of river, has tried to cash in on the mayhem by holding an annual "redneck fishing tournament," which participants are encouraged to attend in costume. "Protective gear is highly recommended!" the tournament's website advises.

The day I got hit, I was out on the Illinois River with another group of contract fishermen doing "barrier defense." Also on the trip were several other tagalongs, including a professor named Patrick Mills. Mills teaches at Joliet Junior College, which is just a few miles from the spot where the Corps is hoping to erect its "disco" noise-and-water-jet barrier. "Joliet is kind of the tip of the spear," he told me. He was wearing a Joliet Junior College baseball cap with a GoPro camera clipped to the bill.

Mills was one of several people I met in Illinois who, for reasons that were not always entirely clear to me, had decided to throw themselves into the fight against Asian carp. A chemist by training, he'd developed a special kind of flavored bait that was supposed to attract carp to the nets. With the help of a local confectioner, he'd produced a truckload of prototypes. These were the size and shape of bricks and made mostly of melted sugar. "It's a bit MacGyvered," Mills acknowledged.

The flavor being tested on this day was garlic. I sampled one of the baits, and it tasted, not unpleasantly, like a garlicky Jolly Rancher. Mills informed me that the following week would be devoted to anise. "Anise is a very good river flavor," he said.

Mills's work had attracted the interest of the U.S. Geological Survey, and a research biologist had come up from Columbia, Missouri—a six-hour drive—to see how the trials were going. The candymaker who'd helped make the baits had come, too, and so had his wife. The Illinois River at this point, about eighty miles from Chicago, was wide and untrafficked. A pair of bald eagles soared overhead, and fish jumped around and sometimes into the boat. Everyone seemed in a festive mood, except for the fishermen, for whom this was, so to speak, just another day at the office.

A few days earlier, the fishermen had set out a couple of dozen hoop nets, which look and function like wind socks. (The nets expand when there's water flowing through them and collapse when there isn't.) Half of the hoop nets had been baited with Mills's bricks, which hung in little mesh bags. The hope was that the baited nets would attract more carp. The fishermen made no secret of their skepticism. One of them griped to me about the smell of the carp candy, a complaint I found curious since the odor it was competing with was the stench of dead fish. Another rolled his eyes at what he saw as a waste of money.

"In my opinion, it's a joke," the most outspoken of the group, Gary Shaw, said to Mills at one point. The sugar dissolved so fast he didn't see how the carp could sense the flavor or find the bait. Mills responded diplomatically. "We have these ideas, but only through these conversations can we improve them," he said. When all the hoop nets had been emptied, the fishermen hauled the catch to another semi-trailer. These fish, too, were destined for fertilizer.

Ideas about how to keep Asian carp out of the Great Lakes can seem as numerous as the carp themselves. "We get calls every day from people," Kevin Irons told me. "We've heard everything—from barges that all the fish jump into to knives flying through the air. Some are more thoughtful than others."

Irons is the assistant chief of fisheries at the Illinois Department of Natural Resources and, as such, he spends most of his working hours worrying about carp. "I hesitate to dismiss any idea too early," he said the first time I spoke to him, over the phone. "You never know which little thought might spark interest."

For his part, Irons believes the best hope for halting the invasion is to enlist what might, with a certain amount of squinting, be seen as a biological agent. What species is large and voracious enough to make a serious dent in the carps' numbers?

"Humans know how to overfish things," Irons told me. "So the question is: How can we use this to our advantage?"

A few years ago, Irons organized an event to encourage people to love carp to death. He called it CarpFest. I attended the inaugural gathering, which was held at a state park not far from Morris. Near the park's boat launch was a huge white tent; inside, volunteers were handing out all manner of invasive-species swag. I picked up a pencil, a refrigerator magnet, a pocket guide titled *Invaders of the Great Lakes*, a hand towel that said FIGHT THE SPREAD OF AQUATIC INVADERS, and a tip sheet for fending off flying carp.

"Clip the 'kill' switch to your clothing," the tip sheet, published by the Illinois Natural History Survey, advised. "This will prevent the boat from continuing its progression if you get knocked out of or thrown from the boat." From a company that turns carp into pet treats, I received a free package of dog chews, which resembled mummified snakes.

I found Irons sitting next to a map showing how Asian carp could use the Sanitary and Ship Canal to slip into Lake Michigan. He's a burly man with sparse white hair and a white beard who looks like Santa might look if Santa, in the off-season, carried a tackle box.

"People feel passionately about the Great Lakes, the ecosystem, even though it's highly altered," he said. "We have to be careful about saying, 'Oh, this pristine system,' because it's not really natural anymore." Irons himself grew up in Ohio, fishing on Lake Erie. In recent years, Lake Erie has been subject to algae blooms that turn huge expanses of the water a nauseous green. Were Asian carp to make their way into Lake Michigan and from there into the other lakes, the blooms, biologists fear, would provide them with an all-you-can-eat buffet. The gorging carp might help cut down on the algae, but, in the process, they'd displace sport fish like walleye and perch.

"Lake Erie, that's where we'd most likely see the greatest impact," Irons said.

As we talked, a large man was cutting up a large silver carp in the center of the tent. A group had gathered around to watch.

"You see, I angle my knife," the man, Clint Carter, explained to the assembled spectators. He had skinned the fish and was now cutting long strips of flesh from its flanks.

"You can take these and grind them and make your fish patties and fish burgers," Carter told the group. "You can't tell the difference between that and a salmon burger."

Of course, in Asia, people have been happily eating Asian carp for centuries. This is the whole reason for raising the "four famous domestic fishes" and, indirectly at least, the reason they came to the attention of American biologists back in the 1960s. A few years ago, when a group of U.S. scientists visited Shanghai to learn more about the fish, the *China Daily* ran an article headlined ASIAN CARP: AMERICANS' POISON, CHINESE PEOPLE'S DELICACY.

"Chinese people have eaten the tasty fish, which are a rich source of nutrition, since ancient times," the paper noted. Accompanying the article were photos of several savory-looking dishes, including milky carp soup and stewed carp with chili sauce. "Serving a carp whole is a symbol of prosperity in Chinese culture," the paper said. "At a banquet it is customary to serve the whole fish last."

China is an obvious market for America's Asian carp. The problem, Irons explained to me, is that the fish would have to be frozen for export, and the Chinese prefer to buy their fish fresh. Americans, for their part, are put off by the fishes' boniness. Bighead and silver carp have two rows of what are known as intramuscular bones; these are shaped like the letter Y and make it all but impossible to produce a bone-free fillet.

"People hear Asian carp—'carp' is a four-letter word—and they're like 'ewww,'" Irons said. But then, when they try it, they change their tune. One year, Irons recalled, the Illinois DNR served carp-based corn dogs at the state fair: "Everybody loved them."

Carter, who owns a fish market in Springfield, is, like Irons, a carp-eating evangelist. He told me that one of his friends had his nose broken by a jumping carp and, as a result, had to have eye surgery.

"We need to control them," he said. "If you can catch millions and tens of millions of pounds of them, it's going to help, and the only way to do that is to create a demand for them." He took the strips he'd cut, rolled them in breadcrumbs, and deep-fried them. It was a warm late-summer day, and by this point he was sweating profusely. When the strips were done, he offered them around as samples, to general approval.

"Tastes like chicken," I heard one boy say.

Sometime around noon, a man in a white chef's jacket showed up at the tent. Everyone referred to him as Chef Philippe, though his full name is Philippe Parola. Parola, originally from Paris, now lives in Baton Rouge, and he'd made the trip to northern Illinois—twelve hours by car, though Parola said he'd done it in ten—to promote his own idea of a killer dish.

Parola was smoking a fat cigar. He handed around more swag—T-shirts that showed a carp smoking a fat cigar and eyeing a frying pan with alarm. SAVE OUR RIVERS, the shirts read on the back. He'd also brought along a large box. On one side of the box was printed THE ASIAN CARP SOLUTION and, under that, CAN'T BEAT 'EM, EAT 'EM! Inside were fish cakes that resembled giant meatballs.

"With a little spinach bed, a little cream sauce, this can be an appetizer," Parola said in a thick French accent, as he passed around a plate of the cakes. "You put two of these with fries, with cocktail sauce, that can be served at a football stadium. You can put them on a tray for a wedding reception. So the diversity of the product is unbelievable."

Parola told me he'd devoted nearly a decade of his life to devising his cakes. Much of that time he'd spent banging his head against the Y-bone problem. He'd tried specialized enzymes and high-tech deboning machines imported from Iceland; the only result was Asian carp mush. "Every time I was trying to cook something with it, it was turning gray, and it tasted like pastrami," he recalled. Finally, he concluded that the fish would have to be deboned by hand, but, since labor costs in the United States were prohibitively high, he would need to outsource.

The cakes he'd brought to CarpFest had been made from fish caught in Louisiana. These had been frozen and shipped to Ho Chi Minh City. There, Parola related, the carp had been thawed, processed, vacuum-packed, refrozen, and put on another container ship, bound for New Orleans. In a concession to Americans' anti-carp prejudice, he'd rechristened the fish "silverfin," a term he'd had trademarked.

It was hard to know how many miles Parola's "silverfin" had

traveled in their journey from fingerlings to finger food, but I figured it had to be at least twenty thousand. And that wasn't counting the trip their ancestors had made to get to the United States in the first place. Did this really represent "the Asian Carp Solution"? I had my doubts. Still, when the cakes came my way, I took two of them. They were, indeed, quite tasty.

2

New Orleans Lakefront Airport sits on a tongue of fill that sticks out into Lake Pontchartrain. Its terminal is a splendid Art Deco affair that at the time of its construction, in 1934, was considered state of the art. Today, the terminal is rented out for weddings and the tarmac used for small planes, which is how I came to be there, a few months after CarpFest, riding shotgun in a four-seat Piper Warrior.

The Piper's owner and pilot was a semi-retired lawyer who liked having an excuse to fly. Often, he told me, he volunteered his services to transport rescue animals between shelters. Dogs, he indicated, without quite saying so, were his favorite passengers.

The Piper took off to the north, over the lake, before looping back toward New Orleans. We picked up the Mississippi at English Turn, the sharp bend that brings the river almost full circle. Then we continued to follow the water as it wound its way into Plaquemines Parish.

Plaquemines is the southeasternmost tip of Louisiana. It's where the great funnel of the Mississippi basin narrows to a spout and Chicago's flotsam and jetsam finally spill out to sea. On maps, the parish appears as a thick, muscular arm thrust into the Gulf of Mexico, with the river running, like a vein, down its center. At the very end of the arm, the Mississippi divides into three, an arrangement that calls to mind fingers or claws, hence the area's name—the Bird's Foot.

Seen from the air, the parish has a very different look. If it's an arm, it's a horribly emaciated one. For most of its length—more than sixty miles—it's practically all vein. What little solid land there is clings to the river in two skinny strips.

Flying at an altitude of two thousand feet, I could make out the houses and farms and refineries that fill the strips, though not the people who live or work in them. Beyond was open water and patchy marsh. In many spots, the patches were crisscrossed with channels. Presumably, these had been dug when the land was firmer, to get at the oil underneath. In some places, I could see the outlines of what were once fields and now are rectilinear lakes. Great white clouds, billowing above the plane, were mirrored in the black pools below.

Plaquemines has the distinction—a dubious one, at best—of being among the fastest-disappearing places on earth. Everyone who lives in the parish—and fewer and fewer people do—can point to some stretch of water that used to have a house or a hunting camp on it. This is true even of teenagers. A few years ago, the National Oceanic and Atmospheric Administration officially retired thirty-one Plaquemines place names, including

Bay Jacquin and Dry Cypress Bayou, because there was no there there anymore.

And what's happening to Plaquemines is happening all along the coast. Since the 1930s, Louisiana has shrunk by more than two thousand square miles. If Delaware or Rhode Island had lost that much territory, America would have only forty-nine states. Every hour and a half, Louisiana sheds another football field's worth of land. Every few minutes, it drops a tennis court's worth. On maps, the state may still resemble a boot. Really, though, at this point, the bottom of the boot is in tatters, missing not just a sole but also its heel and a good part of its instep.

A variety of factors are driving the "land-loss crisis," as it's come to be called. But the essential one is a marvel of engineering. What leaping carp are to Chicagoland, sunken fields are to the parishes around New Orleans—evidence of a man-made natural disaster. Thousands of miles of levees, flood walls, and revetments have been erected to manage the Mississippi. As the Army Corps of Engineers once boasted: "We harnessed it, straightened it, regularized it, shackled it." This vast system, built to keep southern Louisiana dry, is the very reason the region is disintegrating, coming apart like an old shoe.

And so a new round of public-works projects is under way. If control is the problem, then, by the logic of the Anthropocene, still more control must be the solution.

Start to dig in Plaquemines or almost anywhere in southern Louisiana and you will turn up peaty mud; the consistency of the region's soil has been compared to warm Jell-O. Pretty soon, your hole will fill with water. This makes it hard to keep things like caskets underground, which is why the dead in New Orleans are stored in vaults. Keep digging and eventually you'll hit sand and clay. Dig on and you will reach more sand and more clay, and this process will repeat for hundreds—in some places thousands—of feet. Except for those that have been imported to shore up the levees and reinforce the roads, there are no rocks in southern Louisiana.

The layers of sand and clay are, in a manner of speaking, imported, too. A version of the Mississippi has been flowing for millions of years, and all the while it has carried on its broad back vast loads of sediment—at the time of the Louisiana Purchase some four hundred million tons' worth annually. "I do not know much about gods; but I think that the river is a strong brown god," T. S. Eliot wrote. Whenever the river overtopped its banks—something it used to do virtually every spring—it cast its sediment across the plain. Season after season, layer after layer, clay and sand and silt built up. In this way, the "strong brown god" assembled the Louisiana coast out of bits and pieces of Illinois and Iowa and Minnesota and Missouri and Arkansas and Kentucky.

Because the Mississippi is always dropping sediment, it's always on the move. As the sediment builds up, it impedes the flow, and so the river goes in search of faster routes to the sea. Its most dramatic leaps are called "avulsions." Over the last seven thousand years, the river has avulsed six times, and each time it has set about laying down a new bulge of land. Lafourche Parish is what's left of the lobe laid down during the reign of Charlemagne. Western Terrebonne Parish is the remains of a delta lobe built during the time of the Phoenicians. The city of New Orleans sits on a lobe—the St. Bernard—created around the time of the Pyramids. Many still-more-ancient lobes are now submerged. The Mississippi fan, an enormous cone of sediment

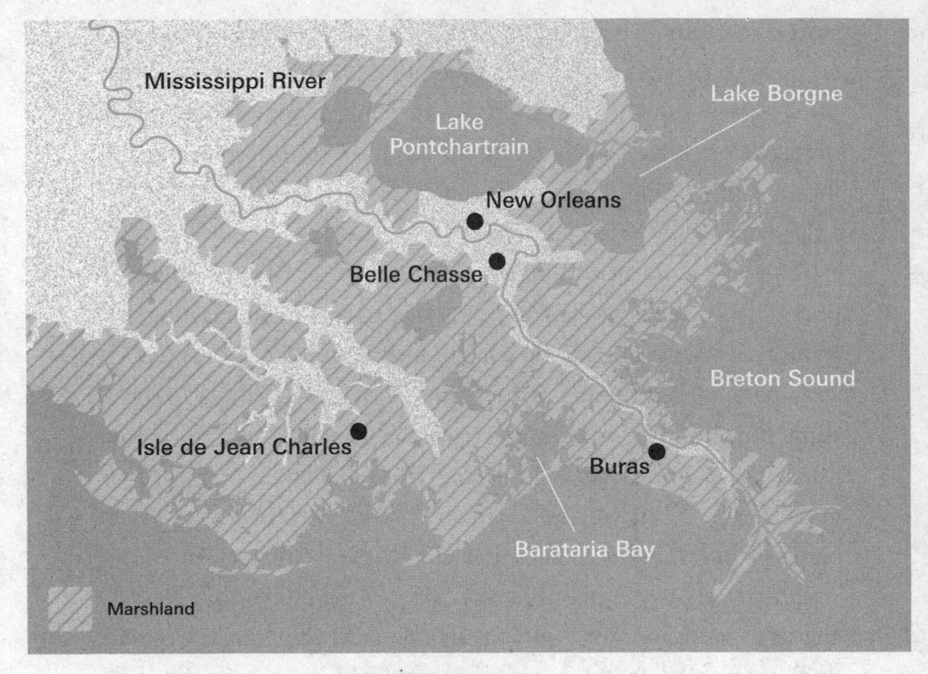

Much of southern Louisiana is no longer dry land.

deposited during the ice ages, now lies under the Gulf; it's larger than the entire state of Louisiana and in some places ten thousand feet thick.

Plaquemines Parish was constructed in this same way. Geologically, it's the baby of the family. It started to form around fifteen hundred years ago, following the river's last great leap. Since it's the youngest lobe, you might think it would be the most long-lasting, but the opposite is true. The delta's soft, Jell-O-like soils tend to dewater and compact over time. The newest layers, which are wetter, lose bulk most rapidly, so as soon as a lobe ceases to grow, it starts to sink. In southern Louisiana, to borrow from Bob Dylan, any place that is "not busy being born is busy dying."

Such a mutable landscape is a tough one to settle. Nevertheless, Native Americans were living in the delta even as it was being created. Their strategy for dealing with the river's vagaries, as far as archaeologists have been able to determine, was one of accommodation. When the Mississippi flooded, they sought higher ground. When it shifted quarters, they did, too.

The French, on arriving at the delta, consulted with the tribes living there. In the winter of 1700, they erected a wooden fort on what's now the east bank of Plaquemines. Pierre Le Moyne d'Iberville, the fort's commander, had been assured by a Bayogoula guide that the site was a dry one. Whether this represented a purposeful misstatement or just a misunderstanding—"dry" in southern Louisiana being a relative term—the place soon flooded out. A priest who visited the following winter found soldiers wading "mid-leg deep" to get to their cabins. In 1707, the fort was abandoned. "I do not see how settlers can be placed on this river," Iberville's brother, Jean-Baptiste Le Moyne de Bienville, wrote to the authorities in Paris, explaining the retreat.

Bienville went on to found New Orleans in 1718, in spite of his cold, wet feet. The new city was called, in honor of its watery surroundings, L'Isle de la Nouvelle Orléans. Not surprisingly, the French chose to build where the land was highest. Counterintuitively, this was right up against the Mississippi, on ridges built by the river itself. During floods, sand and other heavy particles tend to settle out of the water first, creating what are known as natural levees. (*Levée* in French simply means "raised.")

One year after its founding, L'Isle de la Nouvelle Orléans suffered its first inundation. "The site is drowned under half a foot of water," Bienville wrote. The settlement would remain submerged for six months. Rather than retreat again, the French dug in. They raised artificial levees atop the natural ones and started cutting drainage channels through the muck. Most of this backbreaking labor was performed by African slaves. By the

1730s, slave-built levees stretched along both banks of the Mississippi for a distance of nearly fifty miles.

These early levees, made of earth reinforced with timber, failed frequently. But they established a pattern that endures to this day. Since the city wasn't going to move to suit the river, the river would have to be made to stay put. With each flood, the levees were improved—built higher and wider and longer. By the War of 1812, they extended for more than a hundred and fifty miles.

A few days after I flew over Plaquemines, I found myself once again gazing down on the parish. The Mississippi was rising rapidly, and there was concern that the gates on a spillway upriver from New Orleans weren't functioning. If the water kept rising and the spillway failed to open, the city and the parishes downriver from it would be inundated. I was with several engineers, and they were starting to get nervous. I was anxious, too, though only a little, since the Mississippi we were looking at was about five inches wide.

The Center for River Studies is an outpost of Louisiana State University. It's situated not far from the actual Mississippi, in Baton Rouge, in a building that resembles a hockey rink.

At the center's center is a 1:6,000 replica of the delta, from the town of Donaldsonville, in Ascension Parish, to the tip of the Bird's Foot. The model is made from high-density foam that's been machined to mimic the region's topography and all that's been added to it—the levees, the spillways, the floodwalls. The size of two basketball courts, it's sturdy enough to stand on. But when the model is running, as it was the day I showed up, it's hard to take more than a few steps. There are large puddles representing Lake Pontchartrain and Lake Borgne, which are not really lakes but, rather, brackish lagoons. More puddles represent Barataria Bay and Breton Sound, inlets of the Gulf, and still more puddles represent various bayous and backwaters. I pulled off my shoes and tried to walk from New Orleans to the coast. By the time I got to English Turn, my feet were wet. I stuffed my soggy socks into my pocket.

The model delta, which represents a kind of relief map of the future, is supposed to simulate land loss and sea-level rise and to help test strategies for dealing with them. Prominently displayed on one of the walls of the center is a maxim attributed to Albert Einstein: "We cannot solve our problems with the same thinking we used when we created them."

At the time of my visit, the model was so new that it was still being calibrated. This involved running simulations of well-documented disasters from the past, like the flood of 2011. In the spring of that year, heavy snowmelt, along with weeks of intense rain across the Midwest, resulted in record-breaking water levels. To spare New Orleans, the Army Corps of Engineers opened the Bonnet Carré Spillway, about thirty miles upriver from the city. (The Bonnet Carré diverts water into Lake Pontchartrain; when all the gates are open, the flow through it exceeds that of Niagara Falls.) On the model, the spillway gates were represented by small strips of brass attached to copper wires. Because in previous trials the gates had jammed, an engineer had been positioned to watch over them from a folding chair. He looked like a latter-day Gulliver, bent over a drowning Lilliput. He, too, I noticed, had wet socks.

In the world of the model, time as well as space contracts. On its accelerated schedule, a year passes in an hour, a month in five minutes. As I watched, the weeks raced by and the river kept rising. Much to the engineers' relief, this time the gates on the pint-sized Bonnet Carré opened. Water began flowing out of the Mississippi into the spillway, and New Orleans was saved, at least for now.

Two separate vats served as the source for the mini-Mississippi. One provided clear water. The other held the mud of the Little Muddy, though not real mud. This was simulated sediment, imported from France and composed of exactingly milled plastic pellets—teensy, half-millimeter-wide pellets for large grains of sand and even teensier pellets to represent finer particles. The sediment was jet black and stood out against the foam riverbed and surrounding terrain, which were painted bright white.

During the mock flood, some of the pellets had been flushed down the spillway, into Lake Pontchartrain. Others had settled in the riverbed, where they'd formed miniature shoals and sandbars. Most had swooshed past New Orleans and around English Turn. So thick with simulated sediment were the channels of the Bird's Foot that they seemed filled with ink. This inky mix was streaming in dark eddies toward the Gulf, where, had it been real sediment, it would have vanished off the continental shelf.

Here, in black and white, was Louisiana's land-loss dilemma. In the days before floodgates and spillways, a super-wet spring like that of 2011 would have sent the Mississippi and its distributaries surging over their banks. The floodwaters would have wreaked havoc, but they would have spread tens of millions of tons of sand and clay across thousands of square miles of countryside. The new sediment would have formed a fresh layer of soil and, in this way, countered subsidence.

Thanks to the intervention of the engineers, there had been no spillover, no havoc, and hence no land-building. The future of southern Louisiana had instead washed out to sea.

LSU's model Mississippi re-creates the river in miniature.

• • •

Directly next door to the Center for River Studies sits the headquarters of Louisiana's Coastal Protection and Restoration Authority. CPRA was founded in 2005, a few months after Hurricane Katrina hit, inundating New Orleans and leaving more than eighteen hundred people dead. The authority's official mission is to implement "projects relative to the protection, conservation, enhancement, and restoration of the coastal area of the state," which is a nice way of saying it's supposed to prevent the region from disappearing.

One day while I was in Baton Rouge, I met up with two engineers from CPRA at the model. As we chatted, someone flipped a switch controlling projectors in the ceiling. Suddenly, the fields of Plaquemines turned green and the Gulf blue. A satellite image of New Orleans glowed in the crook between the Mississippi and Lake Pontchartrain. The effect was dazzling, if also a little unnerving, as when Dorothy steps out of sepia-toned Kansas into Oz.

"You can see there's not a lot of land in Plaquemines," one of the engineers, Rudy Simoneaux, observed. He was wearing a shirt embroidered with the CPRA emblem, a circle with marsh grass on one side, waves on the other, and a black floodwall in between. "It's kind of frightening when you look at this model and realize how close we all are to water."

Simoneaux and his colleague, Brad Barth, were holding a public meeting that evening in Plaquemines, so after we'd admired the mini-Mississippi for a while, we set out for the real thing. Our destination was Buras, a town ten miles north of the Bird's Foot. We reached the parish seat, Belle Chasse, in time to grab po'boys for lunch. Then we continued south on State Route 23, the only through-road on the parish's west bank. We passed a Phillips 66 refinery, a citrus nursery, and fields as flat and green as pool tables.

Much of Plaquemines lies below sea level—six feet under, people sometimes say. This arrangement is made possible by levees—four sets of them. Two run along the river, one on each bank. Another two—known as "back levees"—run between the parish and the Gulf, to prevent the sea from rolling in. The levees, which keep water out, also keep water in. When they are breached or overtopped, Plaquemines fills up like a pair of long, skinny bathtubs.

Plaquemines was devastated by Katrina, which made landfall in Buras, and then was ravaged again, just a few weeks later, by Hurricane Rita, the most intense storm ever recorded over the Gulf. For months after these back-to-back disasters, Route 23 was blocked by washed-up fishing boats. Dead cows hung from the trees. In anticipation of the next catastrophe, public buildings in the parish stand on improbable pilings. Where other schools might have a gym or a ground-floor cafeteria, South Plaquemines High has enough empty space to park a fleet of tractor trailers. (The school's mascot is a swirling hurricane.) Many of the homes in the parish have been similarly elevated. One house we passed had been raised to a particularly vertiginous height; Simoneaux estimated its pilings were thirty feet tall.

"That's really getting up there," he observed. We were driving alongside the river, but inside the levees, so for long stretches the Mississippi was invisible. Every so often, a ship would loom into view. From the vantage point of the road, it appeared to be floating not on water but on air, like a zeppelin.

Near the town of Ironton, Simoneaux pulled off the highway onto a gravel drive. We parked and scrambled over some barbed wire onto a scruffy field. It was a steamy day, and the field, dotted with puddles, smelled of rot. Flies buzzed lazily in the thick afternoon air.

The land we were standing on was a project designated BA-39. Simoneaux explained that, like the rest of the delta, BA-39 had come out of the Mississippi, just not in the usual way. "Picture a massive eight-foot drill bit on the bottom of the river," he said. As the drill spun, it had gouged out sand and mud. Enormous diesel-powered pumps had sent this slurry gushing through a steel pipe thirty inches in diameter. The pipe had run for five miles, from the west bank of the Mississippi, over the river levees, under Route 23, across some cattle fields, over the back levees, and finally into a shallow basin of Barataria Bay. There the muck had piled up until bulldozers spread it around.

BA-39 had proved, not that further proof was really necessary, what enough pipes and pumps and diesel fuel can accomplish. Nearly a million cubic yards of sediment had made the five-mile journey, resulting in the creation—or, to be more exact, the re-creation—of one hundred and eighty-six paludal acres. Here were all the benefits of flooding without the messy side effects: drowned citrus groves, drowned people, cows hanging from the trees. "We took centuries of land-building and we did it in a year," Simoneaux observed. The bill for the project had been $6 million, which, I calculated, meant that the acre we were standing on had cost about $30,000. CPRA's somewhat redundantly titled "comprehensive master plan" calls for dozens more such "marsh creation" projects, each with a price tag of millions or, in some cases, tens of millions of dollars. But Louisiana is locked in a race with the Red Queen, and in this race it has to move twice as fast just to stay even. To match the pace of land loss, the state would have to churn out a new BA-39 every nine days. Meanwhile, with the drill removed, the pumps unplugged, and the pipes carted off, the artificial marsh had already begun to dewater and subside. According to the authority's projections, in another decade, BA-39 will once again have sunk away.

We reached Buras at around 3 P.M. and turned in at a sign advertising CAJUN FISHING ADVENTURES. The sign showed ducks and fish leaping into the air as if startled by some kind of explosion. Behind a grove of palms was an A-frame lodge with a pool out back.

Ryan Lambert, a fish-and-game guide and the lodge's owner, came out to greet us. "I want to teach people not to listen to propaganda," he said, explaining why he'd volunteered to host the evening's meeting. "I want them to see for themselves." To this end, he'd also organized a flotilla of boats to take attendees out on the Mississippi. I joined a group that included a reporter from the local Fox News station and Lambert's big black dog.

Out on the water, it was about ten degrees cooler than on

shore. A stiff breeze set the dog's ears flapping like flags. We hit the wake of another boat, and the Fox reporter, trying to balance a camera on his shoulder, almost fell overboard.

Unlike Plaquemines's west bank, where the levees stretch all the way to the Bird's Foot, on the east bank they give out right about at the point where, if the parish were actually an arm, its elbow would go. South of the elbow, the river regularly overflows. On occasion, it cuts a new channel, sending water and sediment flowing in new directions and, in the process, creating new land.

"Everything you see in front of you used to be open water," Lambert said as we glided past a wide stretch of green. "Now it's lush and beautiful." His mirrored sunglasses reflected the low late-afternoon sun and the tea-colored river.

"Look at all the new willow trees!" he exclaimed. He was steering with one hand and gesturing with the other. "Look at the birdlife!" The Fox reporter asked what the spot was called.

"It's hard to put a name on it, because it doesn't have a name, because it's new," Lambert said. "This is the newest land in the world!"

We sped in and out of unnamed bayous. A large alligator sunning itself on a log plopped into the water as we zipped by. "Isn't this beautiful?" Lambert kept saying. "When I come here, I feel great. When I go over to the west bank, I want to vomit." The newborn marsh had the sweet smell of freshly cut grass. Way off in the distance, I could see the silhouette of a giant oil platform, perched above the Gulf.

Back on the west bank, in the lodge, the meeting was about to start. A screen had been set up in a room decorated with an elk's head, a stuffed squirrel, and several fish mounted in splashy poses. About fifty people had gathered, some sitting on couches, others leaning against the walls beneath the elk and the fish.

Barth began with a slide presentation. He explained the region's deep geology—how the coast had built up over the millennia, delta lobe by delta lobe, as the Mississippi thrashed around. Then he laid out the problem: How were two million people going to live in a region that was sinking into oblivion? The losses were particularly acute, he noted, in their own backyard. The area around Plaquemines had already shrunk by some seven hundred square miles.

"We're in an uphill battle against sea-level rise and subsidence," Barth said. CPRA would continue to drill and lay pipe. "We will attempt to dredge every ounce of sediment out of the river that we can," he promised. But projects like BA-39 were incommensurate with the scale of the challenge: "We need to be bold."

When the Mississippi bursts through its levees, be they natural or man-made, the opening is called a "crevasse." For most of New Orleans's history, the term was a synonym for disaster.

In 1735, a crevasse-induced flood inundated practically all of New Orleans, which at that point consisted of forty-four square blocks. Sauvé's Crevasse was a breach that flooded the city in May 1849. A month later, a reporter for *The Daily Picayune*, surveying New Orleans from the cupola of the St. Charles Hotel, observed "one sheet of water, dotted in innumerable spots with houses." In 1858, forty-five crevasses opened up in Louisiana's levees, in 1874, forty-three, and in 1882, two hundred and eighty-four.

In what's become known as the "Great Flood of 1927," two hundred and twenty-six crevasses were reported. That flood inundated twenty-seven thousand square miles across a half-dozen states. It displaced more than half a million people, caused an estimated $500 million worth of damage (more than $7 billion in today's money), and marked a very wet watershed. "I woke up

A contemporary rendering of Sauvé's Crevasse

this mornin', can't even get out of my door," Bessie Smith lamented in "Backwater Blues."

In response to the "great flood," Congress in effect nationalized flood control along the Mississippi and entrusted the work to the Army Corps of Engineers. Joseph Ransdell, Louisiana's senior U.S. senator at the time, called the Flood Control Act of 1928 the most important piece of water-related legislation "since the world began." The Corps extended the levees—within four years, it had added another two hundred and fifty miles' worth—and strengthened them. (On average, the levees were raised by three feet, while their volume almost doubled.) The Corps also added a new feature—spillways, like the Bonnet Carré. When the river was at flood stage, the spillway gates would open, relieving pressure on the levees. A poem commemorating the Corps' efforts declared:

> The plan was an engineer masterpiece
> Fashioned by experts, a grand bas-relief
> Levees, floodways, and other improvements
> Blended into a project beneficent.

Thanks to the "project beneficent," the crevasse period came to an end. But with the end of river flooding came an end to fresh sediment. In the succinct formulation of Donald Davis, a geographer at LSU: "The Mississippi River was controlled; land was lost; the environment changed."

CPRA's "bold" scheme for saving Plaquemines is to rehabilitate the crevasse for a post-crevasse age. The agency's master plan calls for punching eight giant holes through the levees on the Mississippi and two more through those on its main distributary, the Atchafalaya. The openings will be gated and channelized, and the channels will themselves be leveed. CPRA likes to characterize the effort as a form of restoration—as a way to "reestablish the natural sediment deposition process." And this is true, but only in the sense that electrifying a river might be called natural.

The furthest along of the man-made crevasses is a project known as the Mid-Barataria Sediment Diversion. The diversion will be six hundred feet wide and thirty feet deep and lined with enough concrete and riprap to pave over Greenwich Village. It will start on the west bank of the Mississippi, some thirty-five miles upriver of Buras, then, in evident defiance of hydrology, run in a perfectly straight line due west for two and a half miles, to Barataria Bay. When it's operating at maximum capacity, some seventy-five thousand cubic feet of water will pass through it every second. In terms of flow, this will make it the twelfth-largest river in the United States. (For comparison's sake, the Hudson River's average flow is twenty thousand cubic feet per second.) Nothing quite like it has ever been attempted before. "It's one of a kind," Barth told me.

Currently, the bill for the project is estimated at $1.4 billion. The next diversion in line, the Mid-Breton, which is planned for the east bank of Plaquemines, is priced at $800 million. Financing for both diversions is supposed to come out of the settlement fund from the BP oil spill, which, in 2010, spewed more than three million barrels of oil into the Gulf, fouling the coast from Texas to Florida. (Planning for the other eight diversions is still at an early stage and funding for them hasn't yet been secured.)

Many Plaquemines residents, like Lambert, welcome the diversions as the parish's last best hope. "It's all about the sediment," Albertine Kimble, an outspoken proponent of the projects and one of the few people in the parish who live outside the levees, told me. But there are also many who oppose them. A few weeks before the meeting in Buras, Plaquemines's president had staged a public showdown with CPRA by denying the authority permits to take soil samples at the proposed site of the diversion. The authority had taken them anyway, with a state trooper standing guard.

At Cajun Fishing Adventures, Barth clicked through slides showing where the Mid-Barataria Diversion would go and how it would be constructed. An animation of the process revealed it to be almost incomprehensibly complex, involving relocating a rail line, rerouting Route 23, and assembling the enormous gates out of floating sections. Once the structure was completed, Barth explained, it would allow CPRA to simulate flooding. When the river was running high and carrying the most sand, the gates would be opened. Sediment-rich water would rush across Plaquemines into Barataria Bay. After a few years, enough sand and silt would be deposited that terra semi-firma would start to form. The diversion would be powered by the river itself, instead of by pumps. In contrast to projects like BA-39, it would continue to deliver sediment year after year.

"When we talk about a sediment diversion, what's the main purpose?" Barth said. "It's to maximize the sediment and minimize the fresh water."

A man in the corner of the room raised his hand. "I assume you're going to build it," he said of the Mid-Barataria project. "But what is the damage going to be?" Despite Barth's assurances, the man was worried about how much fresh water would be directed into the basin and how that would affect recreational fishing. "Speckled trout will be done," he declared.

"If this were a natural crevasse, I'd be all for it," he said. "But when we as humans intervene, it rarely turns out well. That's why we are where we are today."

Soon it would be too hot.

It was another sticky day and I'd circled back to New Orleans to meet with a coastal geologist named Alex Kolker. Kolker teaches at the Louisiana Universities Marine Consortium and, as a pedagogical sideline, he sometimes organizes bike excursions around the city. In contrast to more conventionally popular tours, which feature ghosts, voodoo, and pirates, his emphasize hydrology. He'd agreed to take me on one, though he'd warned we'd have to leave early. By noon, the streets would be a sauna.

"This city was built largely by the river," Kolker observed as we set out from the Garden District, which was still sound asleep. "The short story is that the high ground is near the river and the low ground is old swamps and old marshes." We pedaled north on Josephine Street, away from the Mississippi and imperceptibly downhill. Lofty mansions gave way to shotgun houses in various states of renewal and disrepair.

Kolker braked at an enormous pothole. It had been patched with asphalt, and this patch had developed a new pothole of its own. "Subsidence happens on a couple of different scales," he observed. "You have the big scale, where the old marshes are degrading. And then you have smaller-scale features, like this." A bit farther on, we came to a manhole cover sticking up out of the street like a turret.

"The manhole is probably anchored so it doesn't sink, or at least it doesn't sink as fast as the ground around it," Kolker explained. A sign nearby read EVACUATION ROUTE.

In the sunny accounts aimed at tourists, New Orleans is called the "Crescent City," for the curve of the river it was built along, or the "Big Easy," for its laid-back vibe. In a less upbeat context, residents refer to it as the "bowl." By now, most of the bowl lies at or below sea level—some spots as much as fifteen feet below. When you're in the city, it's hard to imagine the entire place sinking underneath you, yet it is. A recent study that relied on satellite data found some parts of New Orleans dropping by almost half a foot a decade. "That's one of the fastest rates on earth," Kolker noted.

After a few more stops to admire various swales and depressions—"There's a sinkhole over there!"—we arrived at the Melpomene Pumping Station. By this point we were in Broadmoor, a low-lying neighborhood sometimes called "Floodmoor." The station was locked, but through its windows I could see a series of what looked like rockets resting on their sides. These were Wood Screw Pumps, named for their inventor, A. Baldwin Wood. Wood patented his design in 1920, a moment of particularly grandiloquent confidence in the power of engineering.

"New Orleans' drainage problem is a terrible one," a front-page article in the *Item* observed in May of that year. "To meet the problem, New Orleans has constructed the greatest drainage system in the world.

"Man every day is surpassing Nature," the article declared. "He has thrown back the giant Mississippi and made it go where it listeth not."

In 1920, New Orleans boasted six pumping stations, including the Melpomene. These allowed "the old swamps" to be drained and converted into new communities, like Lakeview and Gentilly. Today there are twenty-four stations, which together operate one hundred and twenty pumps. During a storm, rain is funneled into a Venice's-worth of canals. Then it's channeled into Lake Pontchartrain. Without this system, large swaths of the city would quickly become uninhabitable.

But New Orleans's world-class drainage system, like its world-class levee system, is a sort of Trojan solution. Since marshy soils compact through dewatering, pumping water out of the ground exacerbates the very problem that needs to be solved. The more water that's pumped, the faster the city sinks. And the more it sinks, the more pumping is required.

"Pumping is a big part of the issue," Kolker told me, as we climbed back onto our sweaty bicycles. "It accelerates subsidence, so it becomes a positive feedback loop."

As we cycled on, the conversation turned to Katrina. Kolker moved to New Orleans about eighteen months after the storm hit. He recalled that for several years, the "bathtub ring"—the citywide stain left behind by the floodwaters—was still clearly visible on the sides of most buildings.

"Here we're getting into areas that had five to eight feet of water," he told me at one point.

An unusually large storm, Katrina was far from a worst-case scenario. As it churned north in the early-morning hours of August 29, 2005, its eye passed to the east of the city. This meant the strongest winds also passed to the east, over towns like Waveland and Pass Christian, in Mississippi. Briefly, it seemed that New Orleans had been spared.

But the storm was driving water into a network of channels along the city's eastern edge. These channels—the Industrial Canal, the Gulf Intercoastal Waterway, and the Mississippi River–Gulf Outlet (popularly known as "Mr. Go")—had been dug for shipping, to provide a shortcut between the river and the sea. Around 7:45 A.M., the levees on the Industrial Canal failed, sending a twenty-foot-high wall of water crashing through the Lower Ninth Ward. At least six dozen people in the predominantly Black neighborhood were killed.

Water was also surging into Lake Pontchartrain. As the hurricane pushed inland, this water was forced south, out of the lake and into the city's drainage canals. The effect was like emptying a swimming pool into a living room. Soon the floodwalls on the 17th Street and London Avenue Canals gave way. By the next day, eighty percent of the bowl was underwater.

Hundreds of thousands of people had evacuated New Orleans ahead of the storm. With the city inundated, it was unclear when they would return, or if they should. THE CASE AGAINST REBUILDING THE SUNKEN CITY OF NEW ORLEANS, ran a headline in *Slate* a week after the hurricane.

"It is time to face up to some geological realities and start a carefully planned deconstruction of New Orleans," declared an op-ed in *The Washington Post.* As a temporary fix, the op-ed's author, Klaus Jacob, a geophysicist and an expert on risk management, suggested that some of New Orleans could be converted "into a city of boathouses." The Mississippi could then be allowed to flood again, "to fill in the 'bowl' with fresh sediment." (Jacob went on to warn, in 2011, that New York City's subways would flood during a major storm, a prediction fulfilled the following year by Superstorm Sandy.)

An advisory group appointed by New Orleans's mayor recommended that only the highest areas of the city—those along the river and atop the Gentilly and Metairie Ridges—be resettled. A public planning process should then be conducted to determine which low-lying neighborhoods to reoccupy and which to abandon.

Proposals to allow parts of the city to revert to water were floated and then, one by one, rejected. Retreat might make geophysical sense, but politically it was a nonstarter. And so the Corps was charged, yet again, with reinforcing the levees, this time against storm surges from the Gulf. South of the city, the Corps erected the world's largest pumping station, part of a $1.1 billion structure called the West Closure Complex. To the east, it built the Lake Borgne Surge Barrier, a concrete wall nearly two miles long and five and a half feet thick that cost $1.3 billion. The Corps plugged the Mississippi River–Gulf Outlet with a nine-hundred-fifty-foot-wide rock dam and installed massive gates and pumps between the drainage canals and Lake Pontchartrain. The pumps at the foot of the 17th Street Canal were designed to move twelve thousand cubic feet of water per second, a flow greater than the Tiber's.

These pharaonic structures have kept the city dry through several recent storms, and, from a certain perspective, New Orleans now appears substantially better protected than when Katrina hit. But what looks like a defense from one angle can look like a trap from another.

"You must have a replenished coast," Jeff Hebert, a former deputy mayor of New Orleans, told me. "Because as goes the coast, so goes New Orleans." Since the close of the crevasse period, land loss to the south has brought the city some twenty miles closer to the Gulf. It's been estimated that for every three miles a storm has to travel over land, its surge is reduced by a foot. If this is the case, then the threat to New Orleans has grown seven feet higher.

"Drive out nature though you will with a pitchfork," Horace wrote in 20 B.C., "yet she will always hurry back, and before you know it, will break through your perverse disdain in triumph."

Toward the end of our subsidence tour, Kolker and I cycled through the French Quarter, where, though it was still early, tourists armed with drinks were jamming the streets. In Woldenberg Park, we got up on top of the levees and looked out over the Mississippi, toward Algiers.

I asked Kolker how he saw the future. "Sea level will continue to rise," he said. The diversions planned for Plaquemines would add some land back to the marshes south of the city, and so, too, would more-conventional dredging projects, like BA-39. "But I think the areas that don't get restored will flood more and more frequently. There will be continued wetland loss." The city once known as L'Isle de la Nouvelle Orléans would, in coming years, Kolker predicted, "look more and more like an island."

Isle de Jean Charles, in Terrebonne Parish, lies fifty miles southwest of New Orleans and a few decades ahead of it. The island can be reached by a single, narrow causeway, which used to ride over land. Time it right and you can now fish from your car.

"In the springtime, there's always water on the road, whenever there's a south wind," Boyo Billiot told me. We were standing in the backyard of the house he had grown up in, which his mother still occupied. It teetered above us on twelve-foot pilings. Several American flags fluttered from the aerial porch. It was winter and the tail end of deer-hunting season. Billiot was dressed in camouflage. His phone kept dinging with messages from hunting buddies wondering where he was.

Billiot is a broad man with a gravelly voice and a salt-and-pepper goatee. He can trace his ancestry back to Jean Charles

Naquin, who gave the island its name in the early 1800s. (The eponymous Jean Charles was an associate of the pirate Jean Lafitte.) Naquin had a son, Jean Marie, who married a Native woman and escaped to the island after his father disowned him. Jean Marie's children, in turn, married descendants of three tribes: the Biloxi, the Chitimacha, and the Choctaw. Most of their children remained on the island, where they formed their own tightly knit, largely self-sufficient society.

"They went for years and years and nobody knew there was anybody living here," Billiot told me. "When they had the Great Depression, they didn't know anything about it over here, because it didn't affect them."

Billiot grew up on Isle de Jean Charles in the 1950s, speaking a mixture of Cajun French and Choctaw. "Everybody knew each other from one end of the island to the other," he recalled. People still earned a living mostly from fishing, oystering, and trapping. His father had had a shrimp boat that he'd docked right in front of the house. In those days, a deep bayou ran the length of the island, and people crabbed in it. The road, which had just been built, didn't get much use, because the island had its own grocery stores.

Today, the stores are all gone. There are about forty houses left, most of them raised up on pilings and many of them abandoned. Since Billiot was a child, Isle de Jean Charles has shrunk from thirty-five square miles to half a square mile—a loss in area of more than ninety-eight percent.

The island is disappearing for all the usual reasons. It's part of an ancient delta lobe whose soil is compacting. Sea levels are rising. In the early part of the twentieth century, it lost its main sources of fresh sediment to flood-control measures. Then came the oil industry, which dug canals through the wetlands. The canals pulled in salt water, and, as the salinity rose, the reeds and marsh grasses died. The die-off widened the channels, allowing in more salt water, causing more die-off and more widening.

"It's almost like when we used to have video players and you'd hold down the fast-forward button to get where you wanted to in a movie," Billiot's daughter, Chantel Comardelle, told me. She was sitting in the kitchen of the elevated house with Billiot's mother, whom she calls *Maman.* The walls were lined with family photos. "Those canals just held the fast-forward button on the problem."

After back-to-back hurricanes in the 1980s flooded the trailer they were living in, Billiot and Comardelle and the rest of their immediate family moved off the island. With each successive storm, another chunk of land was lost, and more families left. In the early 2000s, a ring of levees was erected around the remnants of Isle de Jean Charles. These turned the bayou where people had once fished and crabbed into a narrow, stagnant pond. Inside the levees, land loss slowed. Outside and along the road, it only got worse.

Even at this point, steps could have been taken to preserve what remained of Isle de Jean Charles. Plans for a massive hurricane-protection system, known as the Morganza to the Gulf project, were being drawn up and could have been extended to include the island. But in this case, the Corps recommended against more engineering. Building the extension would have added $100 million to the project's billion-dollar price tag and preserved just three hundred soupy acres. For that much money, you could buy five times as much land in, say, Chicago.

Residents of the island, as well as the families that have moved off it, are virtually all members of the Isle de Jean Charles Band of the Biloxi-Chitimacha-Choctaw Tribe. Comardelle is the band's secretary, Billiot is a deputy chief, and the band's chief is Billiot's uncle. When it became clear that the road and then, ultimately, the island itself were going to be allowed to wash away, a plan was drawn to move the entire community inland. For the first phase of construction, the band applied for a $50 million federal grant, which was awarded in 2016. At the time of my visit, though, the money had become tied up in state politics, and no one was sure what was going to happen.

As I wandered past empty homes plastered with NO TRESPASSING signs, I could see the economic logic of the island's "planned deconstruction." At the same time, the injustice was pretty glaring. The Biloxi and the Choctaw had come to Louisiana after they'd been dispossessed of ancestral lands, farther east. The Isle de Jean Charles Band had been able to live peacefully on the island only because it was too isolated and commercially irrelevant for anyone else to take an interest in. The band had had no say in the dredging of the oil channels or in the layout of the Morganza to the Gulf project. They'd been excluded from the efforts to control the Mississippi, and now that new forms of control were being imposed to counter the effects of the old, they were being excluded from those, too.

"It's kind of hard to imagine that no one's going to be living here," Billiot told me. "But I've watched it erode away."

From a distance, the Old River Control Auxiliary Structure looks like a row of sphinxes attached at the ears. The structure is four hundred and forty feet long and a hundred feet high. When you get close enough, you can see that the heads of the sphinxes are really cranes and the haunches steel gates. If there's a single feat of engineering that can stand in for the centuries-long effort to dominate the Mississippi—to make it "go where it listeth not"—the Auxiliary Structure might be it. Unlike a levee or a spillway, built to stop the river from flooding, it was put up to stop time.

The Auxiliary Structure sits on a broad plain about eighty miles upriver of Baton Rouge. Near this spot, some five hundred years ago, the Mississippi went on a bender, creating a kind of hydrological, as well as nomenclatural, hairball. The meander took the Mississippi so far west that it ran into the Atchafalaya, at the time a distributary of a different river, the Red, which itself was a Mississippi tributary. The Atchafalaya is a good deal shorter and steeper than the last few hundred miles of the Mississippi, and the tangle presented the water in the larger river with a choice. It could follow its old path to the Gulf, via New Orleans and the Bird's Foot, or it could switch routes and take the faster path offered by the Atchafalaya. Until the mid-1800s, an enormous logjam on the Atchafalaya, which was dense enough to walk across, complicated this choice. But once the jam was removed—by,

The Old River Control Auxiliary Structure

among other means, nitroglycerine—more and more water began flowing out of the main stem of the Mississippi. As the flow on the Atchafalaya increased, it widened and deepened.

In the ordinary course of events, the Atchafalaya would have kept widening and deepening until, eventually, it captured the lower Mississippi entirely. This would have left New Orleans low and dry and rendered the industries that had grown up along the river—the refineries, the grain elevators, the container ports, and the petrochemical plants—essentially worthless. Such an eventuality was thought to be unthinkable, and so, in the 1950s, the Corps stepped in. It dammed the former meander, known as Old River, and dug two huge gated channels. The river's choice would now be dictated for it, its flow maintained as if it were forever the Eisenhower era.

Long before I caught sight of the Auxiliary Structure, I'd read about it in John McPhee's classic piece "Atchafalaya," a morality tale of a darkly comic cast. In McPhee's telling, the Corps throws its heart—and millions of tons of concrete—into forestalling the Mississippi's avulsion and believes it has succeeded.

"The Corps of Engineers can make the Mississippi River go anywhere the Corps directs it to go," one general avers, after a narrow brush with disaster, in 1973, when control of Old River Control was nearly lost. McPhee writes admiringly of the Corps' grit, determination, even genius, but running through the essay is a strong countercurrent. Is the Corps just kidding itself? Are we all?

"Atchafalaya," McPhee writes. "The word will now come to mind more or less in echo of any struggle against natural forces—heroic or venal, rash or well advised—when human beings conscript themselves to fight against the earth, to take what is not given, to rout the destroying enemy, to surround the base of Mt. Olympus demanding and expecting the surrender of the gods."

I showed up at Old River Control on a lovely Sunday afternoon in late winter. The Corps' office, tucked behind a formidable iron fence, looked empty. But when I pressed a buzzer by the driveway, the intercom crackled to life and a resource specialist named Joe Harvey came to the gate. He was dressed as if he was about to go fishing, with his pants tucked into green rubber boots. Harvey led me out to a gazebo overlooking the Auxiliary Structure and its outflow channel.

As the water in the channel swirled by, we chatted about fluvial history. "In 1900, about ten percent of the Red River and the Mississippi put together was going down the Atchafalaya," Harvey explained. "In 1930, you had about twenty percent. By 1950, you had thirty percent." This was the trend line that had prompted the Corps to step in.

"We still do the seventy–thirty division," Harvey said. Every day, engineers measure the flow on the Red and the Mississippi and adjust the gates accordingly. On this particular Sunday, they were allowing through some forty thousand cubic feet per second.

"From here down to the mouth of the Mississippi is about three hundred and fifteen miles," he went on. "And from here to the mouth of Atchafalaya is about a hundred and forty miles. So it's about half the distance. So the river wants to go this way. But if that happens . . ." His voice trailed off.

Two people were fishing the outflow channel from a little motorboat, and I asked Harvey what they might catch. "Oh, we have everything that's in the Mississippi," he said. "Of course, now there's a whole lot of carp, and that's not so good.

"They're still trying to keep them out of the Great Lakes," he added. "Here they're just everywhere."

McPhee included "Atchafalaya" in his book *The Control of Nature*, published in 1989. Since then, a lot has happened to complicate the meaning of "control," not to mention "nature." The Louisiana delta is now often referred to by hydrologists as a "coupled human and natural system," or, for short, a CHANS. It's an ugly term—another nomenclatural hairball—but there's no simple way to talk about the tangle we've created. A Mississippi that's been harnessed, straightened, regularized, and shackled can still exert a godlike force; it's no longer exactly a river, though. It's hard to say who occupies Mount Olympus these days, if anyone.

INTO THE WILD

1

A couple of weeks before the Christmas of 1849, William Lewis Manly climbed to a mountain pass and beheld "the most wonderful picture of grand desolation one could ever see." Manly was standing in what's now southwestern Nevada, not far from Mount Stirling. He imagined his parents, back home in Michigan, with a "bounteous stock of bread and beans" gracing the table, and contrasted this with his own situation—"an empty stomach and a dry and parched throat." The sun was setting as he descended, and his thoughts grew ever gloomier. He began to weep, for, as he would later recall, "I believed I could see the future and the results were bitter to contemplate."

Manly found himself wandering the desert owing to a series

of unfortunate decisions. Three months earlier, he and some five hundred other argonauts had assembled in Salt Lake City, planning to journey together to gold country, in northern California. They'd arrived in Salt Lake too late in the season to take the most direct route, over the Sierras, and so, to avoid getting snowed in, they'd jogged to the south, along a pack trail, toward Los Angeles. A few weeks into the trip, they'd encountered another contingent of forty-niners, led by a fast-talking New Yorker named Orson K. Smith. Smith carried a crude map, which, he claimed, showed a different, faster path west. Most of the members of Manly's group decided to follow Smith, only to reverse course a few days later, when they found their way barred by a canyon so deep it couldn't be crossed by wagon. (Smith himself turned around shortly thereafter.) But Manly and a few dozen others forged ahead, along the illusive shortcut.

The canyon, they soon discovered, was the least of their problems. A detour around it led into some of the most inhospitable terrain on the continent—a rock-strewn waste that probably no white man had ever straggled through before. (A century later, much of the area would be given over to nuclear testing.) Water was scarce and what could be found often was too salty to drink. There was little forage for the oxen, who grew sluggish and emaciated. When one was killed for food, its bones, Manly noted, were filled not with marrow but with a bloody liquid "resembling corruption."

Manly was traveling with a friend who had a wife and three small children. He served as a sort of scout, hiking ahead of the wagons to reconnoiter. The reports he delivered back to camp were so disheartening that after a while his friend asked him please to shut up; his wife couldn't take it anymore. As the party approached Death Valley—at that point an uncharted expanse of desert—the mood grew particularly grim. Sitting around the campfire a few nights after Manly had broken down in tears, one man described the region as the "Creator's dumping place," where he "left the worthless dregs after making a world." Another said it must be "the very place where Lot's wife was turned into a pillar of salt," only the pillar had been "broken up and spread around the country."

Just at the edge of Death Valley, spirits briefly lifted. On a stony ledge, the party chanced upon a cavern that contained a pool of warm, clear water. A few of the men plunged in; one recorded in his diary that he had "enjoyed an extremely refreshing bath." Manly peered into the water and noticed something strange. The pool was surrounded by rock and sand. It was miles from any other water body. Yet it was dancing with fish. Decades later he would remember these tiny "minnows," each "not much more than an inch long."

The cavern the forty-niners chanced upon is now known as Devils Hole and the "minnows" as Devils Hole pupfish, or, scientifically speaking, *Cyprinodon diabolis.* Devils Hole pupfish are, as Manly described them, about an inch long. They are sapphire blue, with intense black eyes and heads that are large for their body size. They're most easily distinguished by an absence; they're missing the pelvic fins that other pupfish possess.

How Devils Hole got its pupfish is, as one ecologist has put it, a "beautiful enigma." The cavern is a geological oddity—a portal to a vast, maze-like aquifer that runs far beneath the ground and holds water left over from the Pleistocene. It seems unlikely that the fish's ancestors could have traveled through the aquifer; the best guess of ichthyologists is that they were washed into Devils Hole at a time when the whole area was wetter. The pool, which is about sixty feet long and eight feet wide, constitutes *Cyprin-*

odon diabolis's entire habitat. This, it's believed, is the smallest range of any vertebrate.

I first learned about Devils Hole thanks to a crime that took place there. On a warm evening in the spring of 2016, three men, all apparently drunk, scaled the chain-link fence that surrounds the cavern. One shot out a security camera, doffed his clothes, went for a dip, and left his underwear floating in the pool. Another vomited. The following day, a single pupfish was found dead, and a necropsy was performed on it. This led to felony charges. The police eventually released surveillance footage, which I watched and watched again. There were jerky shots of the men driving up to the fence in an ATV. Then, from an underwater camera, there were fuzzy shots of two feet walking along a ledge of rock, kicking up bubbles.

Everything about the crime—the piscine necropsy, the county jail's worth of security, the little fish marooned in the middle of the Mojave—intrigued me. I started reading around and happened upon Manly's memoir, *Death Valley in '49*. I learned that desert fish are a rich and diverse group. Every year, the Desert Fishes Council holds a meeting somewhere in northern Mexico or the western United States; typically, the program for the meeting runs to forty pages. Pupfish are so named because males, wrangling over territory, look a bit like puppies tussling. In the Death Valley area alone, there were at one time eleven species and subspecies of pupfish. One is now extinct, another is believed to be extinct, and the rest are all threatened. The Devils Hole pupfish may well be the rarest fish in the world. In an effort to preserve it, a kind of fishy Westworld has been constructed—an exact replica of the actual pool, down to the ledge where the skinny-dipper's feet were caught on tape. Meanwhile, a plume of radioactive water is creeping its way toward the cavern from the

Nevada Test Site. The more I read, the more I thought, I really ought to visit Devils Hole.

Pupfish counts are conducted four times a year at Devils Hole. The counts are made by a team of biologists from the National Park Service, the U.S. Fish and Wildlife Service, and the Nevada Department of Wildlife—agencies that cooperate (and sometimes squabble) over the fish's future. It took me a while to arrange a trip; by then it was time for the summer census and about 105° Fahrenheit.

I met up with the team in the town nearest the cavern—Pahrump, Nevada. Pahrump has one main road, which is lined with fireworks shops, big-box stores, and casinos. From there it's a forty-five-minute drive to Devils Hole, through a mix of desert scrub and emptiness.

In Manly's day, the cavern would have been hard to spot until you practically toppled into it. Today, it's impossible to miss owing to the ten-foot-tall fence, which is topped with barbed wire. One of the biologists had a key that unlocked a gate. This led to a steep, slippery path. Despite the ferocious sun, the bottom of the cavern was in shadow. Even in midsummer, the pool receives only a few hours of direct sunlight each day.

Some of the biologists were lugging pieces of metal scaffolding, which they assembled into a catwalk. Others were toting scuba tanks. Overseeing the whole operation was a Park Service ecologist named Kevin Wilson. Wilson has spent most of his adult life working with *Cyprinodon diabolis* and is regarded as sort of the dean of Devils Hole. (Though Devils Hole is not in Death Valley—it's across the Funeral Mountains, in the Amargosa Valley—for administrative purposes, it's considered part of

Death Valley National Park.) Just before I arrived, Wilson had been featured in an article in the *High Country News* about the aftermath of the break-in. Thanks in good measure to his efforts, the skinny-dipper had ended up in prison. (The vomiter was sentenced to probation.) The reporter had made Wilson out to be a hero—a dogged desert Columbo—but, in the process, she had described him as potbellied and stern. Wilson was still brooding over the description. At one point he turned to the side so I could get a profile view of his stomach.

"Is this a potbelly?" he asked. I suggested it might better be described as a "paunch." Normally Wilson would have been among those preparing to dive, but he'd recently failed some kind of fitness test. This became the subject of more joking.

When all the gear had been transported and assembled, another Park Service biologist, Jeff Goldstein, delivered a safety lecture. Anyone who was injured would have to be helicoptered out, and it could take forty-five minutes or more for a chopper to arrive. "So be careful," he said. Then he took a poll: How many pupfish would turn up?

"I'm thinking a hundred and forty-eight," Wilson guessed. Ambre Chaudoin, also with the Park Service, offered one hundred and forty. Olin Feuerbacher and Jenny Gumm, from Fish and Wildlife, offered, respectively, one hundred and sixty and one hundred and seventy-seven. Brandon Senger, with the state of Nevada, went with one fifty-five. Chaudoin and Feuerbacher, I learned, were married. Feuerbacher told me that he had popped the question at Devils Hole. Wilson made a barfing gesture.

Much like a municipal swimming pool, the pool at Devils Hole has a shallow end and a deep end. The pool's deep end is very deep indeed. According to the Park Service, it descends "over five hundred feet." How much over is a matter of conjecture, since no one has ever touched bottom and lived to tell

A view of Devils Hole looking from the water up

about it. In 1965, two young divers went exploring and never resurfaced. Their bodies are assumed still to be down there, somewhere. At the shallow end is a sloping ledge of limestone, known as "the shelf," which sits about a foot below the surface of the water. It's on the shelf that the fish tend to spawn and also where they find the most food.

Goldstein and Senger, wearing masks, oxygen tanks, shorts, and T-shirts, plunged in. Within a few seconds, they'd vanished into the dark. Meanwhile, Chaudoin, Feuerbacher, and Gumm got down on all fours on the catwalk to count the fish on the shelf. As they called out numbers, Wilson recorded them on a special form.

Once the shelf census was complete, everyone retreated into the shadows to wait for the divers to resurface. Some owlets hidden in a crevice screeched. The sun crept down the western face of the cavern. "Stay hydrated," Wilson admonished. I noticed a bathtub-type ring around the pool and asked Chaudoin about it. She explained it was a function of the pull of the moon; the aquifer beneath us was so massive that it experiences tides.

Though the pupfish inhabit only the pool's upper reaches—they're rarely seen below seventy-five feet—the vastness of the aquifer has nonetheless shaped them. In the desert, the temperature varies dramatically between night and day, winter and summer. The water in the cavern, heated geothermally, maintains a constant year-round temperature of 93°F and a consistent, albeit very low, concentration of dissolved oxygen. The conditions of high temperature and low oxygen should be fatal. Devils Hole pupfish have evolved—somehow—to cope with these conditions and, just as important, only with them. It's believed that the stressfulness of the environment is what caused the fish to lose their pelvic fins; producing the extra appendages just wasn't worth the energy.

Eventually flashes from the divers' headlamps appeared, streaking through the pool like search beams. Goldstein and Senger heaved themselves out of the water. Senger was carrying a dive slate covered with columns of numbers.

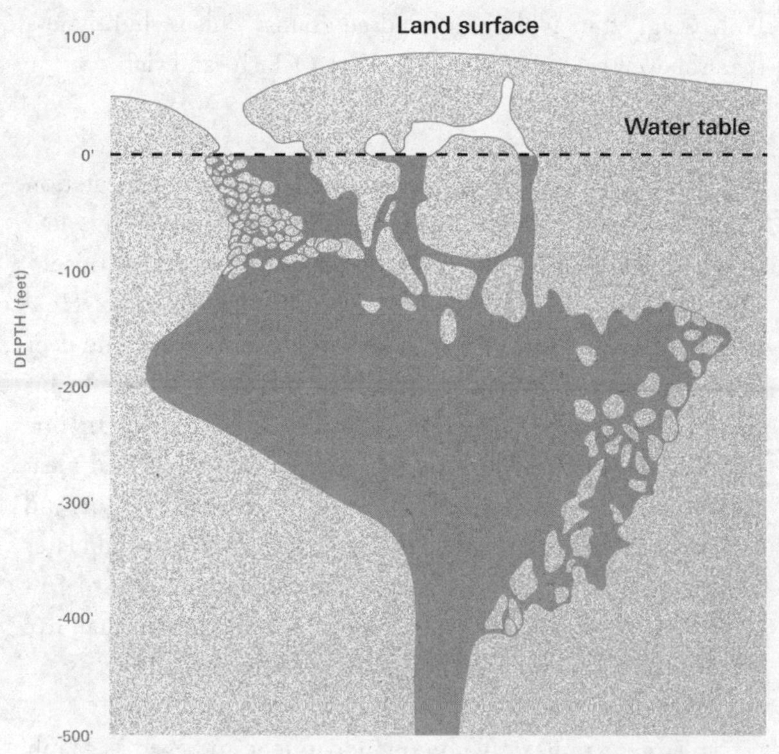

A cross-section view of Devils Hole, showing the canyon in the upper left corner

"That slate holds the key to the universe," Wilson declared.

Everyone climbed back up the rocky path, through the opening in the fence, and out to the parking lot. Senger read off the numbers on the slate. Wilson put these together with the count from the shelf to produce the grand total: one hundred and ninety-five. This was sixty more pupfish than had been counted in the previous census, and higher than anyone had dared guess. High-fives were offered all around. Goldstein did what he called "a little happy dance."

"If there's a lot of fish, we all win," he observed.

Later, I did a calculation. Altogether, the pupfish at Devils Hole weighed in at about a hundred grams. This is slightly less than the weight of a McDonald's Filet-O-Fish sandwich.

When the argonauts set off for the gold fields, the expectation was that a man with steady aim would never starve. Manly had been handed his first rifle when he was fourteen; it was, his father solemnly told him, "suitable for either ball or shot." He'd soon become adept at killing, and the pigeons, turkeys, and deer he bagged were welcome additions to the family's diet. In his early twenties, Manly hunted his way to Wisconsin. In one three-day period, he killed four bears. He ate so much bear meat he spent the next day vomiting. "So long as I had my gun and ammunition I could kill game enough to live on," he would later write. In 1849, he and his companions shot their way to Salt Lake City. An elk Manly brought down weighed more than five hundred pounds and made "the finest kind of food, fit for an epicure."

No larder can be drawn upon indefinitely, and even as Manly was eating his way across the continent, he was helping to make that practice infeasible. In the 1850s, Thoreau lamented the extirpation from New England of moose, cougar, beavers, and wolverines: "Is it not a maimed and imperfect nature that I am conversant with?" Woods that were once thick with wild turkeys were, by the 1860s, all but empty of them. Eastern elk, once plentiful from the Atlantic to the Mississippi, were gone by the 1870s. Passenger pigeons, which formed such immense flocks they blocked the sun, were eliminated around the same time; the last great nesting event—which was also the last great slaughter—took place in 1882.

"It would have been as easy to count or to estimate the number of leaves in a forest as to calculate the number of buffaloes living at any given time during the history of the species previous to 1870," William Hornaday, who served as the chief taxidermist at the Smithsonian and later as director of the Bronx Zoo, wrote. By 1889, Hornaday reckoned, the number of bison living "wild and unprotected" had fallen to fewer than six hundred and fifty. He predicted that in a few years, "hardly a bone will remain above ground to mark the existence of the most prolific mammalian species that ever existed, so far as we know."

Already in Paleolithic times, people had driven plenty of species—woolly mammoths, woolly rhinos, mastodons, glyptodons, and North American camels—into oblivion. Later, as the Polynesians settled the islands of the Pacific, they wiped out creatures like the moa and the moa-nalo. (The latter were goose-like ducks that lived in Hawaii.) When the Europeans reached the islands of the Indian Ocean, they did in, among many other animals, the dodo, the red rail, the Mascarene coot, the Rodrigues solitaire, and the Réunion ibis.

What was different in the nineteenth century was the sheer pace of the violence. If earlier losses had unfolded gradually—so gradually that not even the participants would have been aware of what was going on—the advent of technologies like the railroad and the repeating rifle turned extinction into a readily observable phenomenon. In the United States, and indeed around the world, it became possible to watch creatures vanish in real time. "For one species to mourn the death of another is a new thing under the sun," Aldo Leopold noted in an essay commemorating the passenger pigeon's passing.

In the twentieth century, the biodiversity crisis, as it eventually came to be known, only sped up. Extinction rates are now hundreds—perhaps thousands—of times higher than the so-called background rates that applied over most of geological time. The losses extend across all continents, all oceans, and all taxa. Along with the species formally categorized as endangered, countless others are headed in that direction. American ornithologists have developed a list of "common birds in steep decline"; it includes such familiar creatures as chimney swifts, field sparrows, and herring gulls. Even among insects, a class long thought to be extinction-resistant, numbers are plunging. Whole ecosystems are threatened, and the losses have started to feed on themselves.

As the crow flies, the fake Devils Hole is about a mile from the real one. It's housed in an unmarked hangar-like building, the entrance to which is framed by a pair of signs. One reads CAUTION: PERSONAL PROTECTIVE EQUIPMENT REQUIRED BEYOND THIS POINT, and the second: WARNING! DIHYDROGEN MONOXIDE: USE EXTREME CAUTION.

The first time I visited, I asked about the signs. I was told they'd been put up to deter politically engaged if chemically clueless protesters from trying to break in and trash the place. (Dihydrogen monoxide is a jokey name for water.) Before I was allowed to enter, I had to step into a pail of what looked like urine but turned out to be disinfectant.

Inside, the walls were lined with steel girders, plastic pipes, and electrical wires. A poured-concrete walkway ran around a sunken pool, also made of concrete. The place was about as scenic as a factory floor. In fact, it reminded me of a spent-fuel-rod tank I once saw on a tour of a nuclear power plant. Then again, the fake cavern was fashioned to "bewitch poor fishes' wand'ring eyes," not mine.

Replicating a pool whose bottom has never been touched is clearly impossible, and the deep end of the copy goes down only twenty-two feet. In all other respects, though, it's modeled closely on the original. Since the pool at Devils Hole is almost always in shade, the duplicate has a louvered ceiling that's opened and closed according to the season. Since the water temperature in the cavern is a constant 93°F, at the simulation there's a backup heating system. There's the same shallow shelf, in this case made out of Styrofoam coated with fiberglass, with the same contours. (Laser images of the actual shelf were used to fabricate the replica.)

Not just the pupfish but much of the Devils Hole food chain has been imported into the facsimile. On the Styrofoam shelf float clouds of the same kind of bright-green algae that grow on the limestone version. The water swims with the same species of tiny invertebrates—a spring snail from the genus *Tryonia*, some tiny crustaceans known as copepods, different tiny crustaceans known as ostracods, and a couple of species of beetles.

Conditions in the tank are monitored continuously. If, say, the pH or the water level starts to drop, staff members receive computerized alerts. When major shifts occur, the system sends out phone calls. More than once, Feuerbacher, who works at the facility, has had to drive out from his home in Pahrump in the middle of the night.

Planning for the simulacrum began in 2006. That spring, a bleak one for pupfish, the census hit a record low of thirty-eight. "People were more than a little bit worried about that," Feuerbacher told me. While the $4.5 million facility was under construction, pupfish numbers recovered a bit. Then, in 2013, there was another crash. The spring census yielded just thirty-five pupfish, and the facility, still in the testing phase, was rushed into operation. "We got a call from our higher-ups, saying, 'What's it going to take for you to be ready in three months?'" Feuerbacher recalled.

In the cavern, pupfish live for about a year; in the tank, they can hang on for twice as long. When I visited, Devils Hole Jr. had been in operation for six years. It held about fifty adult fish. Depending on how you look at things, this is a lot of pupfish—fifteen more than the total population on earth in 2013—or not very many. In addition to Feuerbacher, three other people are employed at the facility full-time, which works out to roughly one fishkeeper for every thirteen fish. The number was certainly lower than the Fish and Wildlife Service had hoped for. Feuerbacher thought the explanation might be a beetle.

The beetle, from the genus *Neoclypeodytes*, had been brought over with the other invertebrates from Devils Hole, and it had made the transition to the concrete version all too cheerfully. It was reproducing far faster than in the wild, and somewhere along the way it had developed a taste for pupfish young. One day, Feuerbacher was watching footage from a special infrared camera that's used to capture images of larval pupfish when he saw one of the beetles, which is about the size of a poppy seed, go on the attack.

"It was sort of like a dog catching a scent," he recalled. "It started making tighter and tighter circles around this one larva and then it just dove in and tore it in half." (To extend the dog simile, this would be like a spaniel going after a moose.) In an effort to keep the beetles' numbers in check, the staff had started setting traps for them. Emptying the traps involved sifting their contents through a fine mesh and then picking out each tiny insect with tweezers or a pipette. For an hour or so, I watched two staff members bent over this task, which had to be repeated every day. I was struck, and not for the first time, by how much easier it is to ruin an ecosystem than to run one.

Depending on whom you ask, you'll get a lot of different dates for the onset of the Anthropocene. Stratigraphers, who like clarity, tend to favor the early 1950s. As the United States and the Soviet Union vied for Strangelovian supremacy, aboveground nuclear testing became routine. The tests left behind a more or less permanent marker—a spike in radioactive particles, some of which have a half-life of tens of thousands of years.

Not coincidentally, *Cyprinodon diabolis*'s troubles also date back to this period. In January 1952, President Harry S. Truman added Devils Hole to Death Valley National Park. In a proclamation, Truman said his goal was to protect the "peculiar race of desert fish" that lived in the "remarkable underground pool" and "nowhere else in the world." That spring, the Department of Defense detonated eight nuclear bombs at the Nevada Test Site, about fifty miles north of Devils Hole. The following spring, it detonated eleven more bombs. The mushroom clouds, which were visible from Las Vegas, became a tourist draw.

As the '50s wore on—and more bombs went off—a developer named George Swink started buying up parcels of land around Devils Hole. His plan was to construct from scratch a new town to house test-site workers. Eventually, he bought up some five thousand acres and started to sink wells, including one just eight hundred feet from the cavern.

Swink's scheme stalled, and in the mid-1960s he was bought out by another developer, Francis Cappaert. Cappaert's dream was to make the desert bloom with alfalfa. As soon as he started pumping from the aquifer, the water level in Devils Hole started to drop. By the end of 1969, it had fallen by eight inches. By the

following fall, it had dropped another ten. With each decline, more of the shallow shelf was exposed. By the end of 1970, the pupfish's spawning area had shrunk to the size of a galley kitchen. At this point, a biologist from the University of Nevada came up with the idea of constructing a sham shelf for the fish to breed on. Made out of lumber and Styrofoam, it was installed in the deep end of the pool. Since the deep end receives even less light than the shallow end, the National Park Service rigged up a bank of one-hundred-fifty-watt bulbs to make up the difference. (The fake shelf was eventually destroyed by an earthquake fifteen hundred miles away, in Alaska; because the aquifer is so large, Devils Hole experiences what are known as seismic seiches—in effect, mini-tsunamis.)

Meanwhile, several dozen pupfish were removed from the cavern in an effort to establish backup populations. Some went to Saline Valley, west of Death Valley; others to Grapevine Springs, in Death Valley. A third group was sent to a site near Devils Hole known as Purgatory Spring, and a fourth to a professor at Fresno State, who planned to raise them in an aquarium. All of these early efforts to create a refuge population failed.

By 1972, with more than three-quarters of the shelf exposed, the federal government decided it had no alternative but to sue Cappaert Enterprises. When Truman had set aside Devils Hole, lawyers for the Department of Justice argued, he had also im-

plicitly reserved enough water for the pupfish to survive. The case, *Cappaert v. United States,* would eventually reach the U.S. Supreme Court. As it worked its way through the system, it divided Nevadans. Some saw the fish as an emblem of the desert's fragile beauty. Others saw it as a symbol of government overreach. SAVE THE PUPFISH stickers appeared on car bumpers. Then rival stickers appeared. KILL THE PUPFISH, they said.

Cappaert eventually lost *Cappaert v. United States.* (The fish carried the day nine-to-zero.) In the decades since, his land has been acquired by the Fish and Wildlife Service and converted into the Ash Meadows National Wildlife Refuge. At the refuge, there are some picnic tables, a few trails, and a visitor center that sells, among other items, a plush-toy pupfish that looks like an angry balloon. A pair of signs outside the center note that Cappaert's holdings spanned the ancestral lands of two indigenous peoples: the Nuwuvi and the Newe. In the ladies' room (and perhaps also in the men's), there's a plaque with a passage from Edward Abbey's *Desert Solitaire.* Though the book chronicles Abbey's stint as a ranger in Arches National Park, in Utah, he wrote most of it sitting at a bar in a brothel just a few miles from Devils Hole. "Water, water, water," he observed:

> There is no shortage of water in the desert but exactly the right amount, a perfect ratio of water to rock, of water to sand,

> insuring that wide, free, open, generous spacing among plants and animals, homes and towns and cities, which makes the arid West so different from any other part of the nation. There is no lack of water here, unless you try to establish a city where no city should be.

Jenny Gumm, who manages the fake Devils Hole, has her office in the visitor center, in a part of the building that's off-limits to visitors. One morning, I stopped by to chat with her. A behavioral ecologist by training, Gumm had just moved to Nevada from Texas and was brimming with enthusiasm for her new job.

"Devils Hole is such a special place," she told me. "That experience of going down there, like we did the other day, I've asked people, 'Does this ever get old?' For me it hasn't, and I don't think it will anytime soon."

Gumm pulled out her cell phone. On it was a picture of a pupfish egg. The evening before, one of the staff members at the facility had retrieved the egg from the tank. "There should be a heartbeat by today," she said. "You should be able to see that." The egg, which had been photographed through the eyepiece of a microscope, looked like a glass bead.

Many fish—silver carp, for instance—produce thousands of eggs at a go. This makes it possible to farm them. Devils Hole pupfish release just one pinhead-sized egg at a time. Often these get eaten by the pupfish themselves.

We drove over to Devils Hole Jr. in Gumm's truck and found Feuerbacher in the pupfish nursery—a room filled with rows of glass tanks, assorted equipment, and the burble of running water. Feuerbacher located the egg, which was floating in its own little plastic dish, and put it under the microscope.

When the simulacrum was rushed into operation, in 2013,

one of the first challenges was figuring out how to stock it. With just thirty-five Devils Hole pupfish left on the planet, the National Park Service refused to risk a single breeding pair. It was reluctant even to surrender any eggs. After months of argument and analysis, it finally allowed the Fish and Wildlife Service to gather eggs in the off-season, when the chances of their surviving in the cavern were, in any case, low. The first summer, a single egg was collected; it died. The following winter, forty-two eggs were gathered; twenty-nine of these were successfully reared to adulthood.

The egg under the microscope proved that, beetle problem notwithstanding, the pupfish in the tank were reproducing. It had been collected on a little mat, which had been set out on the fake shelf expressly for this purpose. The mat looked like a piece of tatty shag carpet. "This is a good sign," Gumm said. "Hopefully, there are other eggs that were laid around the carpet that also didn't get eaten."

The egg had, indeed, developed a heartbeat. It had also developed bright-purple swirls—incipient pigment cells. As the tiny heart in the tiny egg pulsated away, I was reminded of the first sonogram images of my own children and of another line from Abbey: "All living things on earth are kindred."

Gumm told me she tries to spend some part of every day by the edge of the tank, just looking at the fish. That afternoon I looked with her. Devils Hole pupfish are, in their own small way, quite flashy. I spotted a pair fooling around, or perhaps flirting, in the deep end. The fish—stripes of blue that seemed almost to glow—circled each other in sinuous unison. Then the pas de deux broke up, and one shot off in an iridescent streak.

"To watch a small school of pupfish arc through a tiny pool of desert water is to discover something vital about wonder," Christopher Norment, an ecologist, wrote after a visit to the real Devils Hole. The same is true, I thought, when the water has been piped in and disinfected. But, I wondered, gazing down at the fish in the tank, wonder about what?

It's often observed that nature—or at least the concept of it—is tangled up in culture. Until there was something that could be set against it—technology, art, consciousness—there was only "nature," and so no real use for the category. It's also probably true that by the time "nature" was invented, culture was already enmeshed in it. Twenty thousand years ago, wolves were domesticated. The result was a new species (or, by some accounts, subspecies) as well as two new categories: the "tame" and the "wild." With the domestication of wheat, around ten thousand years ago, the plant world split. Some plants became "crops" and others "weeds." In the brave new world of the Anthropocene, the divisions keep multiplying.

Consider the "synanthrope." This is an animal that has not been domesticated and yet, for whatever reason, turns out to be peculiarly well suited to life on a farm or in the big city. Synanthropes (from the Greek *syn*, for "together," and *anthropos*, "man") include raccoons, American crows, Norway rats, Asian carp, house mice, and a couple of dozen species of cockroach. Coyotes profit from human disturbance but skirt areas dense with human activity; they have been dubbed "misanthropic synanthropes." In botany, "apophytes" are native plants that thrive when people move in; "anthropophytes" are plants that thrive when people move them around. Anthropophytes can be still further subdivided into "archaeophytes," which were spread before Europeans arrived in the New World, and "kenophytes," which were spread afterward.

Of course, for every species that has prospered with humans, many more have declined, creating the need for another, bleaker list of terms. According to the International Union for Conservation of Nature, which maintains the so-called Red List, a species counts as "vulnerable" when its odds of disappearing within a century are reckoned to be at least one in ten. A species qualifies as "endangered" when its numbers have declined by more than fifty percent over a decade or three generations, whichever is longer. A creature falls into the "critically endangered" category when it's lost more than eighty percent of its population in that same time frame. In IUCN-speak, a plant or animal can be flat-out "extinct," or it can be "extinct in the wild," or it can be "possibly extinct." A species is "possibly extinct" when, on "the balance of evidence," it seems likely to have vanished but its disappearance has not yet been confirmed. Among the hundreds of animals that are currently listed as "possibly extinct" are: the gloomy tube-nosed bat, Miss Waldron's red colobus, Emma's giant rat, and the New Caledonian nightjar. Several species, including the po'ouli, a chubby honeycreeper native to Maui, no longer walk (or hop) the earth but live on as cells preserved in liquid nitrogen. (A term has not yet been coined to describe this peculiar state of suspended animation.)

One way to make sense of the biodiversity crisis would simply be to accept it. The history of life has, after all, been punctuated by extinction events, both big and very, very big. The impact that brought an end to the Cretaceous wiped out something like seventy-five percent of all species on earth. No one wept for them, and, eventually, new species evolved to take their place. But for whatever reason—call it biophilia, call it care for God's creation, call it heart-stopping fear—people are reluctant to be the asteroid. And so we've created another class of animals. These are creatures we've pushed to the edge and then yanked back. The term of art for such creatures is "conservation-reliant," though they might also be called "Stockholm species" for their utter dependence on their persecutors.

The Devils Hole pupfish is a classic Stockholm species. When the water level in the cavern dropped in the late '60s, the sham shelf and the lightbulbs installed by the National Park Service kept the fish alive. After the courts put an end to pumping near the cavern, the water level crept back up, but the aquifer never fully recovered. Today, the water level in the cavern is still about a foot lower than it should be. The ecosystem in the pool has, as a consequence, shifted and the food web frayed. Since 2006, the Park Service has been delivering supplemental meals, including brine shrimp and fairy shrimp—Grubhub for fish.

As for the pupfish in the hundred-thousand-gallon refuge tank, they wouldn't last a season without the ministrations of Gumm, Feuerbacher, and the other fish whisperers. The conditions in the tank are meant to mimic nature as closely as possible, except in the one way that leaves the actual Devils Hole so vulnerable. The simulacrum lies beyond the reach of human disruption because it's totally human.

There is no exact tally of how many species, like the pupfish, are now conservation-reliant. At a minimum, they number in the thousands. As for the forms of assistance they rely on, these, too, are legion. They include, in addition to supplemental feeding and captive breeding: double-clutching, headstarting, enclosures, exclosures, managed burns, chelation, guided migration, hand-pollination, artificial insemination, predator-avoidance training, and conditioned taste aversion. Every year, this list grows. "Old deeds for old people, and new deeds for new," observed Thoreau.

• • •

The Ash Meadows National Wildlife Refuge is twenty-three thousand acres in area, or roughly the size of the Bronx. Within its borders live twenty-six species that can be found nowhere else in the world. According to a brochure I picked up at the visitor center, this represents "the greatest concentration of endemic life in the United States and the second greatest in all of North America."

That harsh conditions should beget diversity is textbook Darwinism. In a desert, populations become physically and then reproductively isolated, much as they do on archipelagoes. The fish of the Mojave and the neighboring Great Basin Desert are, in this sense, like the finches of the Galápagos; each inhabits its own little island of water in a sea of sand.

Doubtless many of these "islands" were sucked dry before anyone bothered to record what was living in them. As Mary Austin observed in 1903, it is the "destiny of every considerable stream in the West to become an irrigating ditch." Among those creatures that lasted long enough for their extinction to be noted were: the Pahranagat spinedace (last collected in 1938), the Las Vegas dace (last seen in 1940), the Ash Meadows poolfish (last seen in 1948), the Raycraft Ranch poolfish (last seen in 1953), and the Tecopa pupfish (missing since 1970).

Another desert pupfish, the Owens pupfish, was thought to be extinct, only to be rediscovered in 1964. By 1969, it was just barely hanging on, in a pond the size of a rec room, when, for reasons no one could quite explain, the pond shrank to a puddle. Someone alerted Phil Pister, a biologist for the California Department of Fish and Game, who rushed to the site—a spot known as Fish Slough. Pister collected all the Owens pupfish left at Fish Slough, with the intention of moving them to a nearby spring. They fit into two buckets.

"I distinctly remember being scared to death," he would later write. "I had walked perhaps fifty yards when I realized that I literally held within my hands the existence of an entire vertebrate species." Pister spent the next several decades working to save the Owens pupfish and also the Devils Hole pupfish. People would often ask him why he spent so much time on such insignificant animals.

"What good are pupfish?" they'd demand.

"What good are you?" Pister would respond.

In the Mojave, I went to see as many fish as I could—island-hopping, as it were. In a pond not far from Devils Hole lives the Ash Meadows Amargosa pupfish (*Cyprinodon nevadensis mionectes*). The pond is surrounded by a landscape so sere it brought to mind Manly's misadventures; just walking the couple of hundred yards from the road, I thought: even today, a person could die in the Mojave and no one would notice. The Ash Meadows pupfish, which look like paler versions of the Devils Hole pupfish, were darting around—once again, either flirting or fighting; I couldn't tell.

Thirty miles away, in the tiny town of Shoshone, California, lives another subspecies, the Shoshone pupfish (*Cyprinodon nevadensis shoshone*). Like the Owens pupfish, the Shoshone pupfish was believed to be extinct, then was rediscovered, in this case in a culvert bordering an RV park. Susan Sorrells owns the RV park, as well as the town's only restaurant and its sole store. With the help of various state agencies, she has created a set of pools for the Shoshone pupfish, which have proved a great deal more adaptable than their Devils Hole cousins.

"They went from being extinct to prolific," Sorrells told me. The hot-springs system that feeds the pupfish ponds also feeds the local swimming pool, which I cooled off in one afternoon along with a bearded man. The man, I was unnerved to see when he turned around, had two large swastikas tattooed on his back.

The town of Pahrump also used to have a fish of its own, the Pahrump poolfish (*Empetrichthys latos*), which still exists, though, sadly, not in Pahrump. The fish's original habitat was a spring-fed pond into which someone, either by design or by chance, released goldfish. The goldfish flourished, while the poolfish crashed. In the '60s, groundwater pumping made a bad situation worse. Just as the pond was about to dry up entirely, in 1971, a University of Nevada biologist named Jim Deacon staged a last-minute rescue. Like Pister, he carried the remaining fish out in a pail. He managed to save thirty-two of them—or at least so the story goes.

Since its rescue, the Pahrump poolfish has lived on in an aquatic diaspora, wandering—or, really, being trucked—from one pond of exile to another. Kevin Guadalupe, a biologist with the Nevada Department of Wildlife, is the fish's Moses. I met up with him at his office, in Las Vegas, which was decorated with a poster showing Nevada's forty species of native fish. "Just about everything on there is endangered," he said, gesturing toward the poster. When he handed me his business card, I noticed it had a pine-nut-sized picture of a poolfish on it.

In the flesh, Pahrump poolfish are about two inches long, with dark, yellow-streaked bodies and yellowish fins. Like Devils Hole pupfish, they evolved in a tough environment where, by default, they were the apex predators. Much of Guadalupe's job involves trying to prevent the poolfish from encountering anything like a real predator. As people move more species into the desert, new emergencies keep arising.

"A lot of the time, we're running around with our hair on fire," Guadalupe told me. At Spring Mountain Ranch, a state park about fifty miles from Pahrump, we visited the shell of a lake that had been home to around ten thousand poolfish. (The ranch once belonged to Howard Hughes, though by the time he bought it he was too paranoid about germs to leave his hotel suite in Las Vegas.) People had dumped the contents of their aquariums into the lake, and, unable to cope with the resulting predation, the poolfish had practically been eliminated. In an effort to get rid of the other introduced species—the poolfish were, of course, themselves transplants—the lake had been completely drained. Its red-clay bottom now lay cracked and baking in the sun. As the environmental historian J. R. McNeill has observed, paraphrasing Marx: "Men make their own biosphere, but they do not make it just as they please."

At Desert National Wildlife Refuge, about forty miles from Pahrump, we toured another pond under siege.

"There's one over there," Guadalupe said, pointing to what looked like a small lobster poking its head out from under some muck. It was a red swamp crayfish. Red swamp crayfish are native to the Gulf Coast, from Mexico to the Florida panhandle. They've been moved around a lot because people like to eat them. For their part, red swamp crayfish like to eat poolfish. To give the fish a chance, Guadalupe had rigged up fake reefs for them to spawn on. These were made of sleek plastic cylinders with tufts of artificial grass sticking out of the top. Guadalupe was hoping that the cylinders would be too slippery for any hungry crayfish to climb.

The last poolfish refuge we hit was in a park in Las Vegas. By the time we got there, it was around noon and a million degrees and no one in their right mind was outside.

That night, my last in Nevada, I stayed on the Strip, at the Paris, in a room with a view of the Eiffel Tower. This being Vegas, the tower rose out of a swimming pool. The water was the blue of antifreeze. From somewhere near the pool, a sound system pumped out a beat that reached me, dull and throbbing, through the sealed windows of the seventh floor. I really wanted a drink. But I couldn't bring myself to go back down to the lobby, past Le Concierge, Les Toilets, and La Réception, to find a faux French bar. I thought of the Devils Hole pupfish in their simulated cavern. I wondered: is this how they felt in their darker moments?

2

Ruth Gates fell in love with the ocean while watching TV. When she was in elementary school, she would sit in front of *The Undersea World of Jacques Cousteau*, mesmerized. The colors, the shapes, the diversity of survival strategies—life beneath the waves seemed to her more spectacular than life above. Without knowing much beyond what she'd learned from the series, she decided that she would become a marine biologist.

"Even though Cousteau was coming through the television, he unveiled the oceans in a way that nobody else had been able to," she told me.

Gates, who grew up in England, went on to study at Newcastle University, where marine-science classes were taught against the backdrop of the North Sea. She took a course on corals and, once again, was dazzled. Her professor explained that corals, which are tiny animals, have, living inside their cells, even tinier plants. Gates wondered how such an arrangement was possible. "I couldn't quite get my head around the idea," she said. In 1985, she moved to Jamaica to study corals and their symbionts.

It was an exciting moment to be doing such work. New techniques in molecular biology were making it possible to look at life at its most intimate level. But it was also a disturbing time. Reefs in the Caribbean were dying. Some were being done in by development, others by overfishing and pollution. Two of the region's dominant reef-builders—staghorn coral and elkhorn coral—were being devastated by an ailment that became known as white-band disease. (Both are now classified as critically endangered.) Over the course of the 1980s, something like half of the Caribbean's coral cover disappeared.

Gates continued her research at UCLA and then at the University of Hawaii. All the while, the outlook for reefs was growing grimmer. Climate change was pushing ocean temperatures beyond many species' tolerance. In 1998, a so-called global bleaching event, caused by a spike in water temperatures, killed more than fifteen percent of corals worldwide. Another global bleaching event took place in 2010. Then, in 2014, a marine heat wave set in and didn't let up for almost three years.

Compounding the dangers of warming were profound changes in ocean chemistry. Corals thrive in alkaline waters, but fossil-fuel emissions were making the seas more acidic. One team of researchers calculated that a few more decades of rising emissions would cause reefs to "stop growing and begin dissolving." Another group predicted that, by the middle of the twenty-first century, visitors to places like the Great Barrier Reef would find nothing more than "rapidly eroding rubble banks." Gates couldn't bring herself to go back to Jamaica; so much of what she loved about the place had been lost.

But Gates was, by her own description, a "glass half full sort of person." She noticed that some reefs that had been given up for dead were bouncing back. These included reefs she knew intimately. What if there were qualities that made certain corals more robust than others? And what if those traits could be identified? Then, perhaps, there'd be something for a marine biologist to do, besides just wring her hands. If it were possible to breed hardier corals, it might be possible to reengineer the world's reefs to survive acidification and climate change.

Gates wrote up her idea and submitted it to a contest called the Ocean Challenge. She won. The prize money—$10,000—was barely enough to keep a research lab in pipettes, but the foundation that sponsored the contest invited her to submit a more detailed proposal. This time around, she received a $4 million grant. News stories about the grant suggested that Gates and her colleagues were planning to create "super coral." Gates embraced the concept. One of her graduate students drew up a logo: a branching coral with a big red S on what might, anthropocentrically, be called its chest.

I met Gates in the spring of 2016. This was about a year after she'd received the super-coral grant and, as it happened, not long after she'd been appointed director of the Hawaii Institute of Marine Biology. The institute occupies its own little island, Moku o Lo'e, in Kaneohe Bay, off the coast of Oahu. (If you've ever watched *Gilligan's Island*, you've seen Moku o Lo'e in the opening sequence.) There's no public transportation to Moku o Lo'e; visitors just show up at a dock, and provided the institute's boatman is expecting them, he'll motor over.

Gates greeted me when I disembarked, and we walked to her office, which was very spare and very white. Its windows looked out over the bay and, beyond it, to a military base—Marine Corps Base Hawaii. (The base was bombed by the Japanese a few minutes before the attack on Pearl Harbor.) Gates explained that Kaneohe Bay had been the inspiration for the super-coral project. For much of the twentieth century, it had been used as a dump for sewage. By the 1970s, its reefs had mostly collapsed. Seaweed had taken over, and the water in the bay had turned an eerily bright green. But then a sewage-treatment plant came online. Later, the state teamed up with the Nature Conservancy and the University of Hawaii to devise a contraption—basically, a barge equipped with giant vacuum hoses—to suck algae off the seabed. Gradually, the reefs started to revive. There are now more than fifty so-called patch reefs in the bay.

"Kaneohe Bay is a great example of a highly disturbed setting where individuals persisted," Gates said. "If you think about the coral that survived, those are the most robust genotypes. So that means what doesn't kill you makes you stronger."

I ended up spending a week with Gates on Moku o Lo'e. One day, we looked at corals through an enormous laser-scanning microscope. Gates showed me the arrangement that as a student she'd found so puzzling. I could see, nestled inside the coral's tiny cells, their tinier plant symbionts. Another day, we went snorkeling. It was two years into the marine heat wave that had begun in 2014, and many of the coral colonies in the bay were a ghostly white. Most of them, Gates observed, probably wouldn't make it. But others were still colorful—tan or brown or greenish. These corals were doing fine. "It's really heartening to see these reefs be so resilient," she told me.

On a third day, we visited an array of outdoor tanks in which corals gathered from the bay were being raised under precisely controlled conditions. The aim wasn't to provide an optimal environment, as at the pupfish tank, but more or less the reverse: the corals were being raised under calibrated stress. Those that thrived—or at least survived—would be crossbred and their offspring thrown back into the tanks for more stress. The corals subject to this selective pressure would, it was hoped, undergo a kind of "assisted evolution." These corals could then be used to seed the reefs of the future.

"I'm a realist," Gates told me at one point. "I cannot continue to hope that our planet is not going to change radically. It already *is* changed." People could either "assist" corals in coping with the change they'd brought about, or they could watch them die. Anything else, in her view, was wishful thinking. "A lot of people want to go back to something," she said. "They think, if we just stop doing things, maybe the reef will come back to what it was.

"Really what I am is a futurist," she said at another point. "Our project is acknowledging that a future is coming where nature is no longer fully natural."

Gates was so charismatic that even though I'd come to Moku o Lo'e with a notebook full of doubts, I felt inspired by her. A couple of times, after she had finished for the day at the institute, we went out for dinner, and eventually we talked our way past the relationship of a reporter and her subject to something approaching friendship. I was arranging to visit Gates again, to see how the super corals were coming along, when she wrote to tell me she was dying. Only she didn't put it that way. Instead, she said that she had lesions on her brain, that she was going to Mexico for treatment, and that, whatever the disease was, she was going to beat it.

Like Ruth Gates, Charles Darwin was confounded by coral. His first encounter with a reef was in 1835. He was sailing on the *Beagle* from the Galápagos to Tahiti when, from the ship's deck, he spied "curious rings of coral land" sticking out of the open sea—what today would be called atolls. Darwin knew that corals were animals and that reefs were their handiwork. Still, the formations baffled him. "These low hollow coral islands bear no proportion to the vast ocean out of which they abruptly rise," he wrote. How, he wondered, was such an arrangement possible?

Darwin cogitated for years on this mystery, which became the subject of his first major scientific work, *The Structure and Distribution of Coral Reefs.* The explanation he came up with—controversial at the time, but now understood to be correct—is that at the center of every atoll lay an extinct volcano. Corals had attached themselves to the volcano's flanks, and as the volcano expired and slowly sank away, the reef had kept growing upward, toward the light. An atoll, Darwin observed, was a kind of a monument to a lost island, "raised by myriads of tiny architects."

The same month that Darwin published his monograph on reefs—May 1842—he sketched out for the first time his revolutionary ideas about evolution, or "transmutation," as the phenomenon was referred to in his day. The sketch was written in pencil and, in the words of one of his biographers, amounted to "thirty-five folio pages of crabbed, elliptical scrawl." Darwin stuck the essay in a drawer. In 1844, he expanded it to two hundred and thirty pages, only once again to hide the manuscript away. There were all sorts of reasons for his reluctance to go public with his ideas, one of which was an almost total lack of evidence.

Darwin was convinced that evolution was unobservable. The process occurred too gradually to be perceived over the course of one human lifetime, or even several. "We see nothing of these slow changes in progress, until the hand of time has marked the lapse of ages," he would eventually write. So how could he prove his theory?

The solution he lit upon was pigeons. In Victorian England, fancy pigeons were a big deal. (Queen Victoria herself kept fancy pigeons.) There were fancy-pigeon clubs, fancy-pigeon shows, and fancy-pigeon poems. "Beneath this laurel's friendly pitying shade/The patriarch of the cote to rest is laid," began an ode to a favorite bird, dead at age twelve. Fanciers fancied dozens of varieties, including: fantails, which, as the name suggests, have extravagant, fan-shaped tail-feather arrangements; tumblers, which, in flight, perform backflips; Nuns, which look like they're wearing ruffs; Barbs, which have a sort of wattle around their eyes; and pouters, which, when they inflate their crops, appear to have swallowed balloons.

A pouter inflating its crop

Darwin set up an aviary in his backyard and used his birds

to perform all sorts of experimental crosses—Nuns with tumblers, for example, and Barbs with fantails. He boiled down the birds' carcasses to get at their skeletons—a task, he reported, that made him "retch awfully." When he finally decided to publish *On the Origin of Species*, in 1859, pigeons strutted across its pages.

"I have kept every breed which I could purchase or obtain," he reports in the opening chapter. "I have associated with several eminent fanciers, and have been permitted to join two of the London Pigeon Clubs."

To Darwin, Nuns and fantails and tumblers and Barbs provided crucial, albeit indirect, support for transmutation. Simply by choosing which birds could reproduce, pigeon breeders had developed lineages that barely resembled one another. "If feeble man can do [so] much by his powers of artificial selection," there was, Darwin speculated, "no limit to the amount of change" that could be effected by "nature's power of selection."

A century and a half after *On the Origin of Species*, Darwin's argument-by-analogy is still compelling, though every year it gets harder to keep the terms straight. "Feeble man" is changing the climate, and this is exerting strong selective pressure. So are myriad other forms of "global change": deforestation, habitat fragmentation, introduced predators, introduced pathogens, light pollution, air pollution, water pollution, herbicides, insecticides, and rodenticides. What do you call natural selection after *The End of Nature*?

Madeleine van Oppen met Ruth Gates at a conference in Mexico in 2005. Van Oppen is Dutch, but at that point she'd been living for almost a decade in Australia. The women were temperamentally opposites—Van Oppen is as reserved as Gates was outgoing; nevertheless, they hit it off immediately. Van Oppen, too, had begun her scientific career as new molecular tools were becoming available, and she, too, had quickly recognized their power. The two began to speak regularly across the time zones and teamed up to write a few papers. Then, in 2011, Gates invited Van Oppen to a conference in Santa Barbara. While there, they realized they'd both become interested in the mechanisms corals use to cope with environmental stress. Could these somehow be harnessed to help them deal with climate change?

"We chatted a lot about this idea of 'assisted evolution,'" Van Oppen told me. "We sort of came up with that term." The application Gates submitted to the Ocean Challenge was written jointly with Van Oppen. It stipulated that, if they won, half the funds would go to Hawaii and half to Australia.

I went to visit Van Oppen almost a year to the day after Gates's death. We met at her office, at the University of Melbourne, which is situated in what used to be the university's botany building, down the hall from a stained-glass window depicting native orchids. The conversation quickly turned to Gates.

"She was so much fun, so full of energy," Van Oppen said. Her face darkened. "It's still unbelievable to me that she's gone. It really makes you realize how fragile life is."

Since I'd been to Hawaii, the super-coral project had progressed, and so, too, had the coral crisis. The heat wave that began in Hawaii in 2014 reached the Great Barrier Reef in 2016, producing another global bleaching event. By the time it ended, the following year, more than ninety percent of the Great Barrier Reef had been affected and something like half its corals had perished. Fast-growing species were particularly hard-hit; they suffered what researchers termed a "catastrophic" collapse. Terry

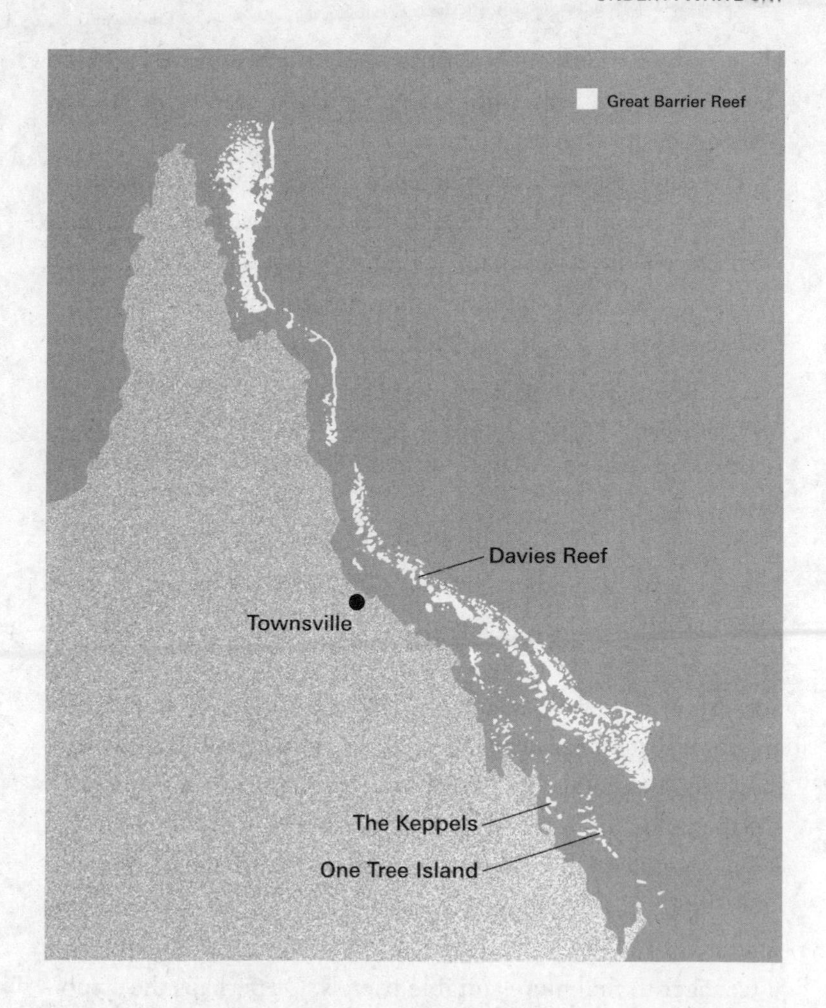

Hughes, a coral biologist at Australia's James Cook University, took an aerial survey of the damage and showed it to his students. "And then we wept," he tweeted.

In a bleaching event, it's the corals' relationship with their symbionts that breaks down. As water temperatures rise, the algae go into overdrive and begin to give off dangerous levels of oxygen radicals. To protect themselves, the corals expel their algae and, as a consequence, turn white. If a heat wave breaks in time, corals can attract new symbionts and recover. If it's too prolonged, they starve to death.

The day I visited, Van Oppen had a meeting with the students and post-docs in her lab. They hailed from a Security Council's worth of countries—Australia, France, Germany, China, Israel, and New Zealand. Van Oppen went around the table, asking for updates. Mostly people reported on the troubles they were having getting one organism or another to cooperate, and mostly Van Oppen just let them rattle on. "That's weird," she said finally to one post-doc whose difficulties seemed especially inexplicable.

As far as Van Oppen and her team were concerned, no member of the reef community was too small to, potentially, make a difference. Some bacteria associated with corals seem particularly adept at scavenging oxygen radicals; one idea the group was exploring was whether it might be possible, by administering some sort of marine probiotic, to make reefs more bleaching-resistant. The corals' algal symbionts, too, might be manipulated. Of the many different types that exist—there are thousands—some seem to be associated with better heat tolerance. Perhaps it would be possible to coax corals to drop less hardy symbionts and take up with a more robust crowd, the way one might coax a teenager to find more suitable friends. Or perhaps the symbionts could themselves be "assisted." One of Van Oppen's post-docs had spent years raising a symbiont variety known as *Cladocopium goreaui* under the sorts of conditions reefs are expected to encounter in the future. (When he showed his *Cladocopium goreaui* to me, I wanted to be amazed; really, though, they just looked like little clouds of dust floating in a jar.) Presumably, the *Cladocopium goreaui* that had come through this rough treatment possessed genetic variants that enabled them to better cope with heat stress. Perhaps "infecting" corals with these hardier strains would help them withstand higher temperatures.

"All the climate models suggest that extreme heat waves will become annual events by mid- to late-century on most reefs in the world," Van Oppen told me. "Rates of recovery are not going to be fast enough to cope with that. So I do think we need to intervene and help them.

"Hopefully the world will come to its senses soon and actually start to reduce greenhouse gases," she went on. "Or maybe there will be some wonderful technological invention that will solve the problem. Who knows what's going to happen? But we need to buy time. So I see assisted evolution as filling that gap, being a bridge between now and the day when we're really holding down climate change or, hopefully, reversing it."

The National Sea Simulator bills itself as "the most advanced research aquarium in the world." It's situated near the city of Townsville, on Australia's eastern coast, about fifteen hundred miles north of Melbourne. Several members of Van Oppen's team work at the facility. They were planning an experiment in assisted evolution, and so, after visiting Van Oppen's lab, I flew up to Townsville.

It was the middle of November and large swaths of Australia were on fire. The news was filled with stories of last-minute escapes, singed koalas, and a pall of smoke over Sydney that made just breathing the equivalent of a pack-a-day habit. On the drive from the airport, I noticed patches of recently burned ground and a billboard with a picture of a raging inferno. ARE

YOU DISASTER READY? the billboard asked. I passed a zinc refinery, a copper refinery, some mango farms, and a wildlife park that advertised crocodile feedings. Dead wallabies—antipodal roadkill—littered the shoulders of the highway.

The SeaSim sits on a spit of land that juts into the Coral Sea. It would have a lovely ocean view, were it not for its lack of windows. Light at the facility is provided by computer-controlled panels of LEDs, which are programmed to mimic the cycles of the sun and moon. Most of the building is given over to tanks. These are set at waist level, like display cases in a department store. As at Gates's lab on Moku o Lo'e, water conditions at the SeaSim can be controlled to produce calibrated stress. In some tanks, the pH and temperature have been set to simulate conditions in the Coral Sea in 2020. Others simulate the hotter seas of 2050, and others the even grimmer conditions expected by the end of the century.

When I arrived it was late afternoon, and the place was nearly

A colony of *Acropora tenuis,* a common species on the Great Barrier Reef

empty. I spent a while just wandering among the tanks, with my nose practically in the water. Individual corals, known unflatteringly as "polyps," are so small they're difficult to see with the naked eye. Even a chunk of coral the size of a child's fist is home to many thousands of polyps, all of which are connected to each other and form a thin layer of living tissue. (The rigid part of a colony is calcium carbonate, which the corals are constantly secreting.) At the SeaSim, tank after tank was filled with a branching species, *Acropora tenuis*, that grows quickly and so is easier to study. *Acropora tenuis* forms colonies that look like miniature pine forests.

As the sun went down, both inside and outside the SeaSim, more and more people started to arrive. So as not to interfere with the light regime, everyone was wearing special red-tinted headlamps that gave off a lurid glow. This seemed appropriate, since what the crowd had come to watch was, we all hoped, an orgy.

Coral sex is a rare and amazing sight. On the Great Barrier Reef, it takes place once a year, in November or December, shortly after a full moon. During the event, called a mass spawning, billions of polyps release in synchrony tiny, bead-like bundles. These bundles, which contain both sperm and eggs, float to the surface and break apart. Most of the gametes become fish food or simply drift away. The lucky ones meet a gamete of the opposite sex and produce a coral embryo.

Tank-raised corals will, if kept under the right conditions, spawn in sync with their relatives out in the ocean. For Van Oppen's team, the spawning offered a critical opportunity to nudge evolution along. The plan was to catch the captive corals in the act, scoop up the gamete bundles, and then, a bit like pigeon fanciers, pick and choose the couplings. One team was hoping to breed *Acropora tenuis* collected from the warmer, northern part of the reef with *Acropora tenuis* collected from the south. A second team had plans to cross altogether-different species of *Acropora* to create hybrids. Some of the offspring of these unnatural hookups would—so the thinking went—be more resilient than their parents.

That evening, the researchers spent hours hovering over the tanks. "This is going to be a big night," one of the scientists who was standing watch told me. "I can feel it." In the run-up to spawning, each polyp develops a tiny bump, making it seem as if the colony has goose pimples. This is called "setting." As we looked on, a few of the colonies set. Then, perhaps out of modesty, perhaps out of anxiety, they held back. Gradually, people gave up and started to drift off to bed. The SeaSim has dorms for just such late nights, but these were full, so I headed out to the parking lot to drive back to Townsville. Making my way through the dark, I could hear the fruit bats screeching in the trees. The following night, I was assured, would be the big one.

The Great Barrier Reef isn't a reef so much as a collection of reefs—some three thousand in all—that stretches over one hundred thirty-five thousand square miles, an area larger than Italy. If there's a more spectacular place on earth—or collection of places—I'm unaware of it. I once spent a week at a research station on a tiny island toward the southern end of the reef, right along the Tropic of Capricorn. Snorkeling off the island, which is called One Tree, I saw corals in mind-bending varieties: branching, bushy, brain-like, plate-like, shaped like fans and flowers and feathers and fingers. I also saw: sharks, dolphins, manta rays, sea turtles, sea cucumbers, octopuses with startled eyes, giant clams with leering lips, and fish in more colors than dreamt of by Crayola.

The number of species that can be found on a healthy patch of reef is probably greater than can be encountered in a similar amount of space anywhere else on earth, including the Amazon rainforest. Researchers once picked apart a single coral colony and counted more than eight thousand burrowing creatures belonging to more than two hundred species. Using genetic-sequencing techniques, other researchers tallied the number of species they could find of crustaceans alone. In one basketball-sized chunk of coral from the northern end of the Great Barrier Reef, they came up with more than two hundred species—mostly crabs and shrimp—and in a similar-sized chunk from the southern end, they identified almost two hundred and thirty species. It's estimated that, worldwide, reefs are home to between one and nine million species, though the scientists who conducted the crustacean study concluded that even the high-end estimates probably are too low. It is likely, they wrote, that "the diversity of reefs" has been "seriously under-detected."

This diversity is all the more remarkable in light of the surroundings. Coral reefs are found only in a band that extends along the equator, from about thirty degrees north to thirty degrees south latitude. At these latitudes, there's not much mixing between the top and the bottom layers of the water column, and essential nutrients, like nitrogen and phosphorus, are in short supply. (The reason the water in the tropics is often so marvelously clear is that little can survive in it.) How reefs support so much diversity under such austere conditions has long puzzled scientists—a conundrum that's become known as "Darwin's paradox." The best answer anyone has come up with is that reef dwellers have developed the ultimate recycling system: one creature's trash becomes its neighbor's treasure. "In the coral city there is no waste," Richard C. Murphy, a marine biologist who worked with Cousteau, has written. "The by-product of every organism is a resource for another."

Since no one knows how many creatures depend on reefs, no one can say how many would be threatened by their collapse; clearly, though, the number is enormous. It's estimated that one out of every four creatures in the oceans spends at least part of its life on a reef. According to Roger Bradbury, an ecologist at Australian National University, were these structures to disappear, the seas would look a lot like they did in Precambrian times, more than five hundred million years ago, before crustaceans had even evolved. "It will be slimy," he has observed.

The Great Barrier Reef is administered as a national park by the Great Barrier Reef Marine Park Authority, which goes by the awkward acronym GBRMPA (pronounced "gabrumpa"). A few months before my visit to Australia, GBRMPA had issued an "outlook report," something it's required to do every five years. The authority said that the reef's long-term prospects, which it had previously characterized as "poor," had declined to "very poor."

Right around the time GBRMPA issued this grim assessment, the Australian government approved a gigantic new coal mine for a site a few hours south of the SeaSim. The mine, often described as a "mega-mine," is expected to send most of its coal to India via a port—Abbot Point—situated right along the reef. Saving corals and mining more coal are, as many commentators pointed out, activities that are tough to reconcile.

"The world's most insane energy project" was *Rolling Stone*'s assessment.

As it happens, GBRMPA has its headquarters in Townsville, in a half-empty mall. On my second day in the city, I walked over to the mall to speak with the authority's chief scientist, David Wachenfeld.

"If we had strongly acted on climate change thirty years ago, I don't know that we'd be having this conversation," Wachenfeld told me. He was wearing a dark-blue polo shirt embroidered with the symbol of the Australian commonwealth, which features a kangaroo gazing at an emu. "We'd be much more likely to be saying, as long as we protect the marine park, we think the reef will look after itself."

As it was, he said, a more interventionist approach was going to be needed. In concert with various universities and research organizations, GBRMPA was planning to spend at least 100 million Australian dollars (about $70 million in American money) investigating ways it might intercede on the reef's behalf. These included: deploying underwater robots to reseed damaged reefs, developing some kind of ultrathin film to shade reefs, pumping deep water to the surface to provide corals with heat relief, and cloud-brightening. This last possibility would involve spraying tiny droplets of salt water into the air to create a kind of artificial fog. The salty mist would, according to theory at least, encourage the formation of light-colored clouds, which would reflect sunlight back out to space, counteracting global warming.

Wachenfeld told me that the new technologies would probably have to be deployed in tandem, so that, for example, a robot might deliver genetically enhanced larvae to a reef shaded by a thin film or man-made fog. "There's all sorts of just amazingly imaginative innovation," he said.

• • •

That evening, I drove back to the SeaSim. Near the parking lot, I noticed a family of feral pigs rooting around. The synanthropes, all fat and sleek, seemed to be having a grand time of it. Gradually, students and researchers drifted over from the dorms. As the simulated sun set over the simulated sea, the place came alive with red lights, zigzagging through the dimness like fireflies.

Everyone from the previous night had returned. In addition to the teams working with Van Oppen, I recognized a group that was planning to freeze coral gametes as an insurance policy against apocalypse and another group looking to genetically manipulate coral embryos. There were some new faces, too. A team of filmmakers had flown in from Sydney. (If the rest of us were coral voyeurs, the filmmakers, it occurred to me, were pornographers.)

The head of the institute that runs the SeaSim, Paul Hardisty, had come for the show, as well. Hardisty, who's from Canada, is tall and rangy in a cowboy-esque way. I asked him about the reef's future. He was at once gloomy and gung-ho.

"We're not talking about coral gardening here," Hardisty told me. "We're talking about major, industrial-scale—all-of-reef-scale—interventions. So it's a really steep curve, but it's possible—that's what we've concluded—with the best minds in the world, all working together." To aid in the research effort, the SeaSim was going to be expanded; if I came back in a few years, Hardisty said, it would be twice the size.

"It won't be a silver bullet," he continued. "It's going to be a combination of things, combinations of, for instance, cloud-brightening and assisted evolution. We're going to need engineering, because we are looking at fast deployment to make a difference. And we're also going to need to borrow technologies from Big Pharma, because we've got to figure out mass-delivery mechanisms. Maybe—I don't know—we'll use little pellets."

The ruby lights swooped and pitched around us. "It's just absolute hubris and so arrogant to think that we can survive without everything else," Hardisty said. "We come from this planet. Anyway, I'm getting a little philosophical. I'm going to have to go home and watch a hockey game."

As we waited for the corals to get in the mood, there wasn't much to do. Standing around in the dark, I found myself also "getting a little philosophical." Hardisty was right, of course; it *was* hubris to imagine that people could drive the Great Barrier Reef to collapse without suffering any consequences. But wasn't it just another kind of hubris to imagine "all-of-reef-scale interventions?"

When Darwin juxtaposed "artificial" and "natural" selection, there was no question in his mind which was more powerful. Pigeon fanciers had done amazing things, breeding varieties so distinctive that to many they seemed different birds entirely. (All the varieties, from fantails to pouters, were, Darwin realized, descended from a single species, the rock pigeon, *Columba livia.*) Dog fanciers, similarly, had bred up greyhounds and corgis, bulldogs and spaniels. The list went on and on: the ewes in the barn, the pears in the garden, the corn in the crib—all were products of generations of attentive breeding.

But, in the grand scheme of things, artificial selection was just tinkering at the margins. It was natural selection—indifferent, but infinitely patient—that had given rise to life's astonishing diversity. In the final, oft-quoted paragraph of *On the Origin of Species,* Darwin conjures an "entangled bank, clothed with many plants of many kinds, with birds singing on the bushes, with various insects flitting about, and with worms crawling through the damp earth." All of these "elaborately constructed forms, so different from each other, and dependent on each other in so complex a manner," had been produced by the same mindless, inhuman force.

"There is grandeur in this view of life," Darwin reassures his readers, whom he imagines still to be skeptical after four hundred and ninety pages. From the very simplest creatures blundering around in the primordial ooze, "endless forms most beautiful and most wonderful have been, and are being, evolved."

The Great Barrier Reef might be thought of as the ultimate "entangled bank." Tens of millions of years of evolution have gone into its creation, with the result that even a fist-sized piece of it is unfathomably dense with life, crammed with creatures "dependent on each other in so complex a manner" that biologists will probably never fully master the relations. And the reef—today, at least—goes on and on.

Everyone I spoke to in Australia understood that preserving the Great Barrier Reef in all its greatness was beyond what could realistically—or unrealistically—be hoped for. Even settling for a tenth of it would mean shading and robotically seeding an area the size of Switzerland. What was at issue was, at best, a diminished thing—a kind of Okay Barrier Reef.

"If we can extend the life of the reef by twenty, thirty years, that might be just enough for the world to get its act together on emissions, and it might make the difference between having nothing and having some sort of functional reef," Hardisty told me. "I mean, it's really sad that we have to talk like that. But that's where we are now."

The second night I spent at the SeaSim also turned out to be a bust. A few colonies set, only to release what one researcher re-

Spawning corals release bead like bundles of eggs and sperm.

ferred to as a "dribble." And so the following evening, I set out for the SeaSim one more time.

By now I knew what to expect. When the sun went down, the researchers would don their headlamps and circulate among the tanks. If they noticed a coral colony set, they would lift it out of its communal tank and place it in a bucket of its own. That evening, so many colonies of *Acropora tenuis* set that it was difficult to get around. Rows of buckets lined the floor. Some of the colonies were from an area known as The Keppels, in the far south of the Great Barrier Reef; others were from a reef known as Davies Reef, hundreds of miles to the north. In the natural course of events, such distant colonies would have no chance to mate. But the whole point of the experiment was not to leave things to nature.

A post-doc named Kate Quigley was in charge of the couplings and of a team of mostly undergraduate volunteers. She wore her red light around her neck, like a glowing amulet. Quigley had laid out dozens of plastic containers, where, if all went well, the inter-reef crosses would occur. Embryos formed in the containers would, she explained, be transferred to little tanks, where they would be subjected to heat stress. Those that survived would then be inoculated with different symbionts, including some of the lab-evolved strains that I'd seen in Melbourne, and then subjected to still more stress.

"We really want to push them to their limits," Quigley told me. "We're really looking for the best of the best."

During my trip to One Tree, I was lucky enough to take a midnight snorkel through a spawning. The scene resembled a blizzard in the Alps, only upside down. Even in a bucket, spawning is a marvel. First, just a few polyps release their bundles; then the rest follow suit, as if prompted by some secret signal. The bundles rise through the water in defiance of gravity. On the surface, they form a rosy-colored slick.

"This is one of the real miracles of nature," I overheard a scientist on the gene-editing team say, more to himself than to anyone else.

As colony after colony let loose, Quigley marshaled her volunteers. She gave each student a bowl and a fine sieve. With a pipette, she extracted the gamete bundles from the buckets and distributed them among the sieves. Out on the reef, the bundles would break apart in the waves; at the SeaSim, the work of the waves would have to be done by hand. Quigley instructed the students to swish the bundles around until they released their contents. The sperm would drop into the bowls, while the eggs, which are larger, would be caught in the mesh.

The students swirled with grave concentration. The eggs

looked like flecks of pink pepper. The bowls of sperm looked like, well, what you'd expect.

"I can take your sperm if you want," I heard a young woman call out.

"Yes, have a bowl of my sperm," a young man replied.

"This is the only place where it's safe to say that," a third student observed.

Quigley had plotted the crosses she wanted to make in a notebook. Under her supervision, the students mixed sperm and eggs from different parts of the reef. This went on late into the night, until every lonely coral had found a mate.

3

Odin, in Norse mythology, is an extremely powerful god who's also a trickster. He has only one eye, having sacrificed the other for wisdom. Among his many talents, he can wake the dead, calm storms, cure the sick, and blind his enemies. Not infrequently, he transforms himself into an animal; as a snake, he acquires the gift of poetry, which he transfers to people, inadvertently.

The Odin, in Oakland, California, is a company that sells genetic-engineering kits. The company's founder, Josiah Zayner, has a shock of dyed-blond hair, multiple piercings, and a tattoo that urges: CREATE SOMETHING BEAUTIFUL. He holds a PhD in biophysics and is a well-known provocateur. Among his many stunts, he has coaxed his skin to produce a fluorescent protein, ingested a friend's poop in a DIY fecal-matter transplant, and attempted to deactivate one of his genes so he could grow bigger biceps. (This last effort, he acknowledges, failed.) Zayner calls himself a "genetic designer" and has said his goal is to give people access to the resources they need to modify life in their spare time.

The Odin's offerings range from a "Biohack the Planet" shot glass, which costs three bucks, to a "genetic engineering home lab kit," which runs $1,849 and includes a centrifuge, a polymerase chain-reaction machine, and an electrophoresis gel box. I opted for something in between: the "bacterial CRISPR and fluorescent yeast combo kit," which set me back $209. It came in a cardboard box decorated with the company's logo, a twisting tree circled by a double helix. The tree, I believe, is supposed to represent Yggdrasil, whose trunk, in Norse mythology, rises through the center of the cosmos.

Inside the box, I found an assortment of lab tools—pipette tips, petri dishes, disposable gloves—as well as several vials containing *E. coli* and all I'd need to rearrange its genome. The *E. coli* went into the fridge, next to the butter. The other vials went into a bin in the freezer with the ice cream.

Genetic engineering is, by now, middle-aged. The first genetically engineered bacterium was produced in 1973. This was soon followed by a genetically engineered mouse, in 1974, and a genetically engineered tobacco plant, in 1983. The first genetically engineered food approved for human consumption, the Flavr Savr tomato, was licensed in 1994; it proved such a disappointment it went out of production a few years later. Genetically engineered varieties of corn and soy were developed at around the same time; these, in contrast to the Flavr Savr, have, in the United States, become more or less ubiquitous.

In the last decade or so, genetic engineering has undergone its own transformation, thanks to CRISPR. CRISPR is shorthand for a suite of techniques—most of them borrowed from bacteria—that make it vastly easier for researchers and biohackers to manipulate DNA. (The acronym stands for "clustered regularly interspaced short palindromic repeats.") CRISPR allows its users to snip a stretch of DNA and then either disable the affected sequence or replace it with a new one.

The possibilities that follow are pretty much endless. Jennifer Doudna, a professor at the University of California, Berkeley and one of the developers of CRISPR, has put it like this: we now have "a way to rewrite the very molecules of life any way we wish." With CRISPR, biologists have already created, among many, many other living things: ants that can't smell, beagles that grow superhero-like muscles, pigs that resist swine fever, macaques that suffer from sleep disorders, coffee beans that contain no caffeine, salmon that don't lay eggs, mice that don't get fat, and bacteria whose genes contain, in code, Eadweard Muybridge's famous series of photographs showing a racehorse in motion. A few years ago, a Chinese scientist, He Jiankui, announced that he had produced the world's first CRISPR-edited humans—twin baby girls. According to He, the girls' genes had been tweaked to confer resistance to HIV, though whether this is actually the case remains unclear. Shortly after he made the announcement, He was placed under house arrest in Shenzhen.

I have almost no experience in genetics and have not done hands-on lab work since high school. Nevertheless, by following the instructions that came in the box from The Odin, I was able, over the course of a weekend, to create a novel organism. First I grew up a colony of *E. coli* in one of the petri dishes. Then I doused it with the various proteins and bits of designer DNA I'd stored in the freezer. The process swapped out one "letter" of the bacteria's genome, replacing an A (adenine) with a C (cytosine). Thanks to this emendation, my new and improved *E. coli* could, in effect, thumb its nose at streptomycin, a powerful antibiotic. If it felt a little creepy engineering a drug-resistant strain of *E. coli* in my kitchen, there was also a definite sense of achievement. So much so, in fact, that I decided to move on to the second project in the kit: inserting a jellyfish gene into yeast in order to make it glow.

The Australian Animal Health Laboratory, in the city of Geelong, is one of the most advanced high-containment laboratories in the world. It sits behind two sets of gates, the second of which is intended to foil truck bombers, and its poured-concrete walls are thick enough, I was told, to withstand a plane crash. There are five hundred and twenty air-lock doors at the facility and four levels of security. "It's where you'd want to be in the zombie apocalypse," a staff member told me. At the highest security level—Biosafety Level 4—are vials of some of the nastiest animal-borne pathogens on the planet, including Ebola. (The laboratory gets a shout-out in the movie *Contagion.*) Staff members who work in BSL-4 units can't wear their own clothes into the lab and have to shower for at least three minutes before heading home. For their part, the animals at the facility can't leave at all. "Their only way out is through the incinerator," is how one employee put it to me.

Geelong is about an hour southwest of Melbourne. On the same trip that I met with Van Oppen, I paid a visit to the laboratory, which goes by the acronym AAHL (rhymes with "maul"). I'd heard about a gene-editing experiment going on there that intrigued me. As a result of yet another biocontrol effort gone awry, Australia is besieged by a species of giant toad known familiarly as the cane toad. In keeping with the recursive logic of the Anthropocene, researchers at AAHL were hoping to address this disaster with a further round of biocontrol. The plan involved editing the toad's genome using CRISPR.

A biochemist named Mark Tizard, who was in charge of the project, had agreed to show me around. Tizard is a slight man with a fringe of white hair and twinkling blue eyes. Like many of the scientists I met in Australia, he's from somewhere else, in his case London.

Before getting into amphibians, Tizard worked mostly on poultry. Several years ago, he and some colleagues at AAHL inserted a jellyfish gene into a hen. This gene, similar to the one I was planning to plug into my yeast, encodes a fluorescent protein. A chicken in possession of it will, as a consequence, give off an eerie glow under UV light. Next, Tizard figured out a way to insert the fluorescence gene so that it would be passed down to male offspring only. The result is a hen whose chicks can be sexed while they're still in their shells.

Tizard knows that a lot of people are freaked out by genetically modified organisms. They find the idea of eating them repugnant and of releasing them into the world anathema. Though he's no provocateur, he believes, like Zayner, that such people are looking at things all wrong.

"We have chickens that glow green," Tizard told me. "And so we have school groups that come, and when they see the green chicken, you know, some of the kids go, 'Oh, that's really cool. Hey, if I eat that chicken, will I turn green?' And I'm, like, 'You eat chicken already, right? Have you grown feathers and a beak?' "

Anyway, according to Tizard, it's too late in the day to be worried about a few genes here and there. "If you look at a native Australian environment, you see eucalyptus trees, koalas, kookaburras, whatever," he said. "If I look at it, as a scientist, what I'm seeing is multiple copies of the eucalyptus genome, multiple copies of the koala genome, and so on. And these genomes are interacting with each other. Then, all of a sudden, *ploomph*, you put an additional genome in there—the cane toad genome. It was never there before, and its interaction with all these other genomes is catastrophic. It takes other genomes out completely.

"What people are not seeing is that this is already a genetically modified environment," he went on. Invasive species alter the environment by adding entire genomes that don't belong. Genetic engineers, by contrast, alter just a few bits of DNA here and there.

"What we're doing is potentially adding on maybe ten more genes onto the twenty thousand toad genes that shouldn't be there in the first place, and those ten will sabotage the rest and take them out of the system and so restore balance," Tizard said. "The classic thing people say with molecular biology is: Are you playing God? Well, no. We are using our understanding of biological processes to see if we can benefit a system that is in trauma."

Formally known as *Rhinella marina*, cane toads are a splotchy brown, with thick limbs and bumpy skin. Descriptions inevitably emphasize their size. "*Rhinella marina* is an enormous, warty bufonid (true toad)," notes the U.S. Fish and Wildlife Service. "Large individuals sitting on roadways are easily mistaken for boulders," observes the U.S. Geological Survey. The biggest cane toad ever recorded was fifteen inches long and weighed almost six pounds—as much as a chubby chihuahua. A toad named Bette Davis, who lived at the Queensland Museum, in Brisbane, in the 1980s, was nine and a half inches long and almost as wide—about the size of a dinner plate. The toads will eat almost anything they can fit in their oversized mouths, including mice, dog food, and other cane toads.

Cane toads are native to South America, Central America, and the very southernmost tip of Texas. In the mid-1800s, they were imported to the Caribbean. The idea was to enlist the toads in the battle against beetle grubs, which were plaguing the region's cash crop—sugar cane. (Sugar cane, too, is an imported species; it is native to New Guinea.) From the Caribbean, the toads were shipped to Hawaii and from there, to Australia. In 1935, a hundred and two toads were loaded onto a steamer in Honolulu. A hundred and one of them survived the journey and ended up at a research station in sugar-cane country, on Australia's northeast coast. Within a year, they'd produced more than 1.5 million eggs. The resulting toadlets were intentionally released into the region's rivers and ponds.

It's doubtful that the toads ever did the sugar cane much good. Cane grubs perch too high off the ground for a boulder-sized amphibian to reach. This didn't faze the toads. They found plenty else to eat and continued to produce toadlets by the truckload. From a sliver of the Queensland coast, they pushed north, into the Cape York Peninsula, and south, into New South Wales. Sometime in the 1980s, they crossed into the Northern Territory. In 2005, they reached a spot known as Middle Point, in the western part of the Territory, not far from the city of Darwin.

Along the way, something curious happened. In the early phase of the invasion, the toads were advancing at the rate of about six miles a year. A few decades later, they were moving at twelve miles a year. By the time they hit Middle Point, they'd sped up to thirty miles a year. When researchers measured the toads at the invasion front, they found out why. The toads on the front lines had significantly longer legs than the toads back in

Queensland. And this trait was heritable. The *Northern Territory News* played the story on its front page, under the headline SUPER TOAD. Accompanying the article was a doctored photo of a cane toad wearing a cape. "It has invaded the Territory and now the hated cane toad is evolving," the newspaper gasped. Contra Darwin, it seemed, evolution *could* be observed in real time.

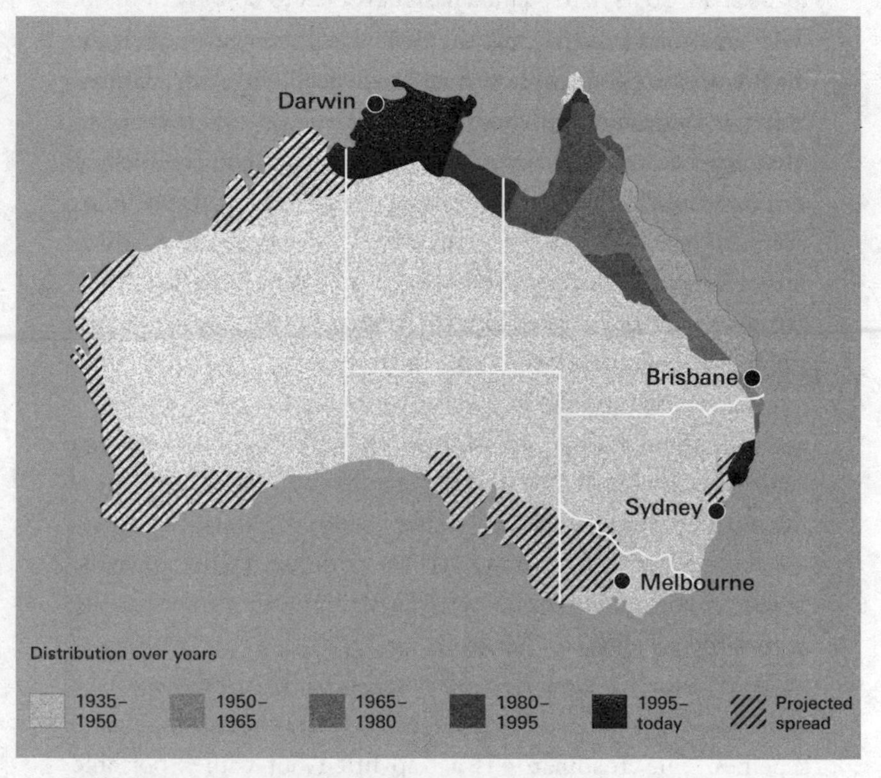

Since they were introduced, cane toads have spread across Australia. They're expected to continue to expand their territory.

Cane toads are not just disturbingly large; from a human perspective, they're also ugly, with bony heads and what looks like a leering expression. The trait that makes them truly "hated," though, is that they're toxic. When an adult is bitten or feels threatened, it releases a milky goo that swims with heart-stopping compounds. Dogs often suffer cane toad poisoning, the symptoms of which range from frothing at the mouth to cardiac arrest. People who are foolish enough to consume cane toads usually wind up dead.

Australia has no poisonous toads of its own; indeed, it has no toads at all. So its native fauna hasn't evolved to be wary of them. The cane toad story is thus the Asian carp story inside out, or maybe upside down. While carp are a problem in the United States because nothing eats them, cane toads are a menace in Australia because just about everything eats them. The list of species whose numbers have crashed due to cane toad consumption is long and varied. It includes: freshwater crocodiles, which Australians call "freshies"; yellow-spotted monitor lizards, which can grow up to five feet long; northern blue-tongued lizards, which are actually skinks; water dragons, which look like small dinosaurs; common death adders, which, as the name suggests, are venomous snakes; and king brown snakes, which are also venomous. By far the most winning animal on the victims list is the northern quoll, a sweet-looking marsupial. Northern quolls are about a foot long, with pointy faces and spotted brown coats. When young quolls graduate from their mother's pouch, she carries them around on her back.

In an effort to slow down the cane toads, Australians have come up with all sorts of ingenious and not-so-ingenious schemes. The Toadinator is a trap fitted out with a portable speaker that plays the cane toad's song, which some compare to a dial tone and others to the thrum of a motor. Researchers at the University of Queensland have developed a bait that can be used to lure cane toad tadpoles to their doom. People shoot cane toads with air rifles, whack them with hammers, bash them with golf clubs, purposefully run them over with their cars, stick them in the freezer until they solidify, and spray them with a compound called HopStop, which, buyers are assured, "anesthetizes toads within seconds" and dispatches them within an hour. Communities organize "toad busting" militias. A group called

An Australian girl with her pet cane toad, Dairy Queen

the Kimberley Toad Busters has recommended that the Australian government offer a bounty for each toad eliminated. The group's motto is: "If everyone was a toad buster the toads would be busted!"

At the point Tizard got interested in cane toads, he'd never actually seen one. Geelong lies in a region—southern Victoria—the toads haven't yet conquered. But one day at a meeting, he was seated next to a molecular biologist who studied the amphibian. She told him that, despite all the busting, the toads kept on spreading.

"She said, it was such a shame, if only there was some new way of getting at it," Tizard recalled. "Well, I sat down and scratched my head.

"I thought: Toxins are generated by metabolic pathways," he went on. "That means enzymes, and enzymes have to have genes to encode them. Well, we have tools that can break genes. Maybe we can break the gene that leads to the toxin."

Tizard brought on a post-doc named Caitlin Cooper to help with the mechanics. Cooper has shoulder-length brown hair and an infectious laugh. (She, too, is from somewhere else—in her case Massachusetts.) No one had ever tried to gene edit a cane toad before, so it was up to Cooper to figure out how to do it. Cane toad eggs, she discovered, had to be washed and then pierced just so, with a very fine pipette, and this had to be done quickly, before they had time to start dividing. "Refining the micro-injection technique took quite a while," she told me.

As sort of a warm-up exercise, Cooper set out to change the cane toad's color. A key pigment gene for toads (and, for that matter, humans) codes for the enzyme tyrosinase, which controls the production of melanin. Disabling this pigment gene should, Cooper reasoned, produce toads that were light-colored instead of dark. She mixed some eggs and sperm in a petri dish, micro-injected the resulting embryos with various CRISPR-related compounds, and waited. Three oddly mottled tadpoles emerged. One of the tadpoles died. The other two—both males—grew into mottled toadlets. They were christened Spot and Blondie. "I was absolutely rapt when this happened," Tizard told me.

Cooper next turned her attention to "breaking" the toads'

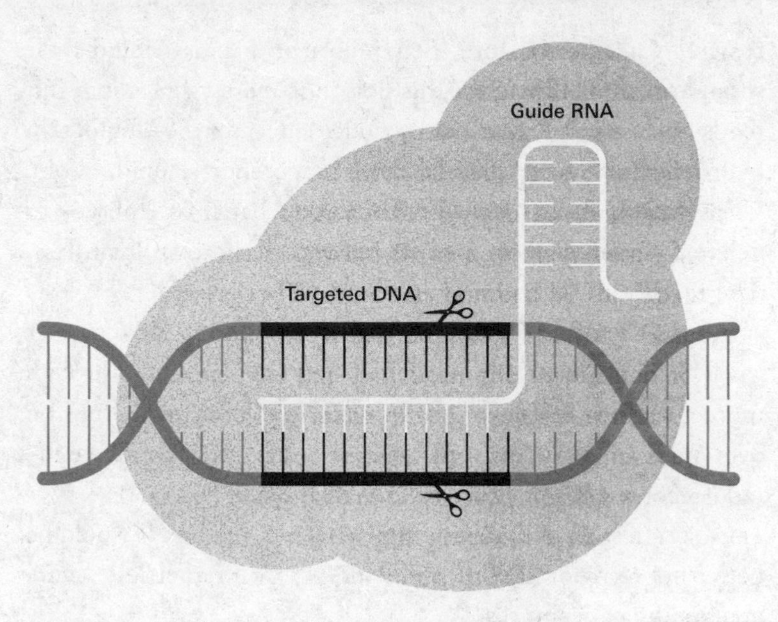

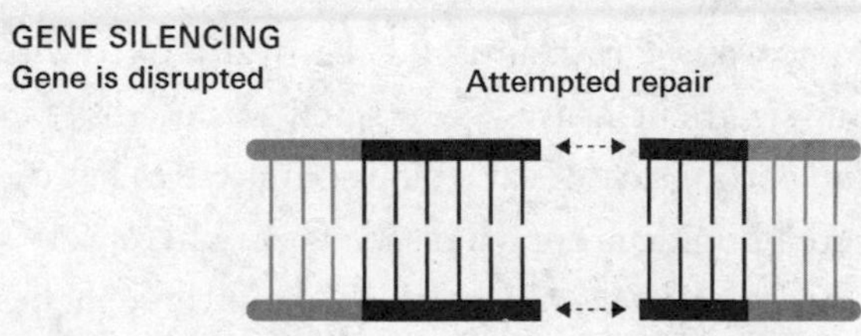

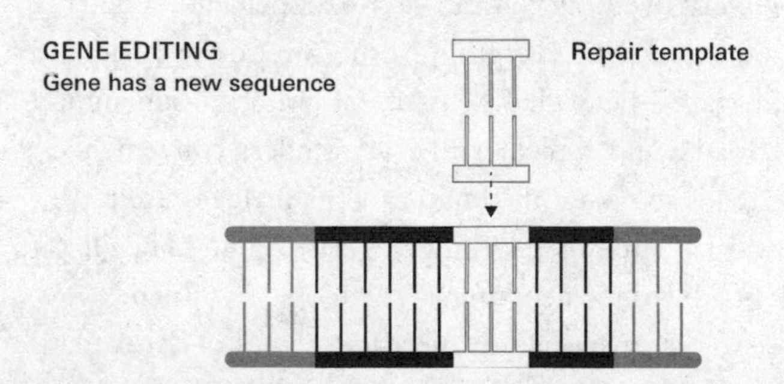

With CRISPR, guide RNA is used to target the stretch of DNA to be cut. When the cell attempts to repair the damage, often mistakes are introduced and the gene is disabled. If a "repair template" is supplied, a new genetic sequence can be introduced.

toxicity. Cane toads store their poison in glands behind their shoulders. In its raw form, this poison is merely sickening. But toads can, when attacked, produce an enzyme—bufotoxin hydrolase—that amplifies the poison's potency a hundredfold. Using CRISPR, Cooper edited a second batch of embryos to delete a section of the gene that codes for bufotoxin hydrolase. The result was a batch of detoxified toadlets.

After we'd talked for a while, Cooper offered to show me her toads. This entailed penetrating deeper into AAHL, through more air-lock doors and layers of security. We all put scrubs on over our clothes and booties over our shoes. Cooper spritzed my tape recorder with some kind of cleaning fluid. QUARANTINE AREA, a sign said. HEAVY PENALTIES APPLY. I decided it would be better not to mention The Odin and my own rather less secure gene-editing adventures.

Beyond the doors was a sort of antiseptic barnyard, filled with animals in various sized enclosures. The smell was a cross between hospital and petting zoo. Near a block of mouse cages, the detox toadlets were hopping around a plastic tank. There were a dozen of them, about ten weeks old and each about three inches long.

"They're very lively, as you can see," Cooper said. The tank had been outfitted with everything a person could imagine a toad would want—fake plants, a tub of water, a sunlamp. I thought of Toad Hall, "replete with every modern convenience." One of the toads stuck out its tongue and nabbed a cricket.

"They will eat literally anything," Tizard said. "They'll eat each other. If a big one encounters a small one, it's lunch."

Let loose in the Australian countryside, a knot of detox toads presumably wouldn't last long. Some would become lunch, either for freshies or lizards or death adders, and the rest would be outbred by the hundreds of millions of toxic toads already hopping across the landscape.

What Tizard had in mind for them was a career in education. Research on quolls suggests that the marsupials can be trained to steer clear of cane toads. Feed them toad "sausages" laced with an emetic, and they will associate toads with nausea and learn to avoid them. Detox toads, according to Tizard, would make an even better training tool: "If they're eaten by a predator, the predator will get sick but not die, and it will go, 'I'm never eating a toad again.'"

Before they could be used for teaching quolls—or for any other purpose—the detox toads would need a variety of government permits. When I visited, Cooper and Tizard hadn't started in on the paperwork, but they were already contemplating other ways to tinker. Cooper thought it might be possible to fiddle with the genes that produce the gel coat on the toads' eggs in such a way that the eggs would be impossible to fertilize.

"When she described the idea to me, I was, like, brilliant!" Tizard said. "If we can take steps to knock down their fecundity, that's absolute gold." (A female cane toad can produce up to thirty thousand eggs at a go.)

A few feet away from the detox toads, Spot and Blondie were sitting in their own tank, an even more elaborate affair, with a picture of a tropical scene propped in front for their enjoyment. They were almost a year old and now fully grown, with thick rolls of flesh around their midsections, like sumo wrestlers. Spot was mostly brown, with one yellowish hind leg; Blondie was more richly variegated, with whitish hind legs and light patches on his forelimbs and chest. Cooper reached a gloved hand into the tank and pulled out Blondie, whom she'd described to me as "beautiful." He immediately peed on her. He appeared to be smiling malevolently, though I realized, of course, that wasn't actually the case. He had, it seemed to me, a face only a genetic engineer could love.

According to the standard version of genetics that kids learn in school, inheritance is a roll of the dice. Let's say a person (or a toad) has received one version of a gene from his mother—call it *A*—and a rival version of this gene—*A1*—from his father. Then any child of his will have even odds of inheriting an *A* or an *A1*, and so on. With each new generation, *A* and *A1* will be passed down according to the laws of probability.

Like much else that's taught in school, this account is only partly true. There are genes that play by the rules and there are also renegades that refuse to. Outlaw genes fix the game in their favor and do so in a variety of devious ways. Some interfere with the replication of a rival gene; others make extra copies of themselves, to increase their odds of being passed down; and still others manipulate the process of meiosis, by which eggs and sperm are formed. Such rule-breaking genes are said to "drive." Even if they confer no fitness advantage—indeed, even if they impose a fitness cost—they're handed on more than half of the time. Some particularly self-serving genes are passed on more than ninety percent of the time. Driving genes have been discovered lurking in a great many creatures, including mosquitoes, flour beetles, and lemmings, and it's believed they could be found in a great many more, if anyone took the trouble to look for them. (It's also true that the most successful driving genes are tough to detect, because they've driven other variants to oblivion.)

Since the 1960s, it's been a dream of biologists to exploit the power of gene drives—to drive the drive, as it were. This dream has now been realized, and then some, thanks to CRISPR.

In bacteria, which might be said to hold the original patent on the technology, CRISPR functions as an immune system. Bacteria that possess a "CRISPR locus" can incorporate snippets of DNA from viruses into their own genomes. They use these snippets, like mug shots, to recognize potential assailants. Then they dispatch CRISPR-associated, or Cas, enzymes, which work like tiny knives. The enzymes slice the invaders' DNA at critical locations, thus disabling them.

Genetic engineers have adapted the CRISPR-Cas system to cut pretty much any DNA sequence they wish. They've also figured out how to induce a damaged sequence to stitch into itself a thread of foreign DNA it's been supplied with. (This is how my *E. coli* were fooled into replacing an adenine with a cytosine.) Since the CRISPR-Cas system is a biological construct, it, too, is encoded in DNA. This turns out to be key to creating a gene drive. Insert the CRISPR-Cas genes into an organism, and the organism can be programmed to perform the task of genetic reprogramming on itself.

In 2015, a group of scientists at Harvard announced they'd used this self-reflexive trick to create a synthetic gene drive in yeast. (Starting with some cream-colored yeast and some red yeast, they produced colonies that, after a few generations, were all red.) This was followed three months later by an announcement from researchers at UC–San Diego that they'd used much the same trick to create a synthetic gene drive in fruit flies. (Fruit flies are normally brown; the drive, pushing a gene for a kind of albinism, yielded offspring that were yellow.) And six months after that, a third group of scientists announced they had created a gene-drive *Anopheles* mosquito.

If CRISPR confers the power to "rewrite the very molecules of life," with a synthetic gene drive, that power increases exponentially. Suppose that the researchers in San Diego had released

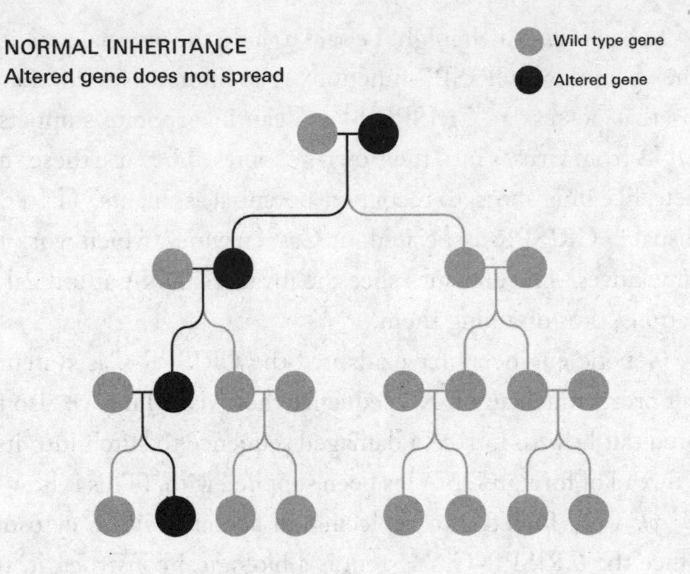

GENE DRIVE INHERITANCE
Altered gene always spreads

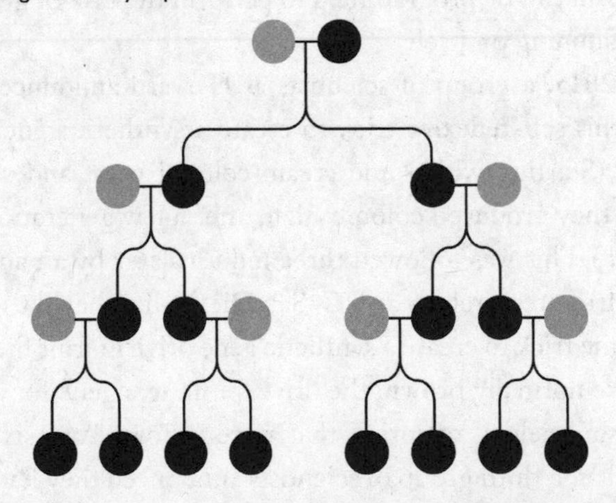

With a synthetic gene drive, the normal rules of heredity are overridden and an altered gene spreads quickly.

their yellow fruit flies. Assuming those flies had found mates, swarming around some campus dumpster, their offspring, too, would have been yellow. And assuming those offspring survived and also successfully mated, their progeny would, in turn, have been yellow. The trait would have continued to spread, ineluctably, from the redwood forest to the Gulf Stream waters, until yellow ruled.

And there's nothing special about color in fruit flies. Just about any gene in any plant or animal can—in principle, at least—be programmed to load the inheritance dice in its favor. This includes genes that have themselves been modified or borrowed from other species. It should be possible, for example, to engineer a drive that would spread a broken-toxin gene among cane toads. It may also be possible one day to create a drive for corals that would push a gene for heat tolerance.

In a world of synthetic gene drives, the border between the human and the natural, between the laboratory and the wild, already deeply blurred, all but dissolves. In such a world, not only do people determine the conditions under which evolution is taking place, people can—again, in principle—determine the outcome.

The first mammal to be fitted out with a CRISPR-assisted gene drive will, almost certainly, be a mouse. Mice are what's known as a "model organism." They breed quickly, are easy to raise, and their genome has been intensively studied.

Paul Thomas is a pioneer in mouse research. His lab is in Adelaide, at the South Australian Health and Medical Research Institute, a sinuous building covered in pointy metal plates. (Adelaideans refer to the building as the "cheese grater"; when I went to visit, I thought it looked more like an ankylosaurus.) As soon as a breakthrough paper on CRISPR was published, in 2012, Thomas recognized it as a game changer. "We jumped on it straightaway," he told me. Within a year, his lab had used CRISPR to engineer a mouse afflicted with epilepsy.

When the first papers on synthetic gene drives came out, Thomas once again plunged in: "Being interested in CRISPR and being interested in mouse genetics, I couldn't resist the opportunity to try to develop the technology." Initially, his goal was just to see if he could get the technology to work. "We didn't really have much funding," he said. "We were doing it on the smell of an oily rag, and these experiments, they're quite expensive."

While Thomas was still, in his words, "just dabbling," he was contacted by a group that calls itself GBIRd. The acronym (pronounced "gee-bird") stands for Genetic Biocontrol of Invasive Rodents, and the group's ethos might be described as Dr. Moreau joins Friends of the Earth.

"Like you, we want to preserve our world for generations to come," GBIRd's website says. "There is hope." The site features a picture of an albatross chick gazing lovingly at its mother.

GBIRd wanted Thomas's help designing a very particular kind of mouse drive—a so-called "suppression drive." A suppression drive is designed to defeat natural selection entirely. Its purpose is to spread a trait so deleterious that it can wipe out a population. Researchers in Britain have already engineered a suppression drive for *Anopheles gambiae* mosquitoes, which carry malaria. Their goal is eventually to release such mosquitoes in Africa.

Thomas told me there were various ways to go about designing a self-suppressing mouse, most having to do with sex. He was particularly keen on the idea of an "X-shredder" mouse.

Mice, like other mammals, have two sex-determining chromosomes—XXs are female, XYs male. Mice sperm carry a single chromosome, either an X or a Y. An X-shredder mouse is a mouse who's been gene edited so that all of his X-bearing sperm are defective.

"Half the sperm drop out of the sperm pool, if you like," Thomas explained. "They can't develop anymore. That leaves you with just Y-bearing sperm, so you get all male progeny." Put the shredding instructions on the Y chromosome and the mouse's offspring will, in turn, produce only sons, and so on. With each generation, the sex imbalance will grow, until there are no females left to reproduce.

Thomas explained that work on a gene-drive mouse was going slower than he'd hoped. Still, he thought by the end of the decade someone would develop one. It might be an X-shredder, or it might rely on a design that's yet to be imagined. Mathematical modeling suggests that an effective suppression drive would be extremely efficient; a hundred gene-drive mice released on an island could take a population of fifty thousand ordinary mice down to zero within a few years.

"So that's quite striking," Thomas said. "It's something to aim for."

If the Anthropocene's clearest geological marker is a spike in radioactive particles, its clearest biological marker may be a spike in rodents. Everywhere humans have settled on the planet—and even some places they've only visited—mice and rats have tagged along, often with ugly consequences.

The Pacific rat (*Rattus exulans*) was once confined to Southeast Asia. Starting about three thousand years ago, seafaring

Polynesians carried it to nearly every island in the Pacific. Its arrival set off wave after wave of destruction that claimed at least a thousand species of island birds. Later, European colonists brought to those same islands—and to many others—ship rats (*Rattus rattus*), thus setting off further waves of extinctions, which are still ongoing. In the case of New Zealand's Big South Cape Island, ship rats didn't arrive until the 1960s, by which point naturalists were on hand to document the results. Despite intensive efforts to save them, three species endemic to the island—one bat and two birds—disappeared.

The house mouse (*Mus musculus*) originated on the Indian subcontinent; it can now be found from the tropics to very near the poles. According to Lee Silver, author of *Mouse Genetics*, "Only humans are as adaptable (some would say less so)." Under the right circumstances, mice can be just as fierce as rats, and every bit as deadly. Gough Island, which lies more or less midway between Africa and South America, is home to the world's last two thousand pairs of Tristan albatrosses. Video cameras installed on the island have recorded gangs of *Mus musculus* attacking albatross chicks and eating them alive. "Working on Gough Island is like working in an ornithological trauma center," Alex Bond, a British conservation biologist, has written.

For the last few decades, the weapon of choice against invasive rodents has been Brodifacoum, an anticoagulant that induces internal hemorrhaging. Brodifacoum can be incorporated into bait and then dispensed from feeders, or it can be spread by hand, or dropped from the air. (First you ship a species around the world, then you poison it from helicopters!) Hundreds of uninhabited islands have been de-moused and de-ratted in this way, and such campaigns have helped bring scores of species back from the edge, including New Zealand's Campbell Island teal, a small, flightless duck, and the Antiguan racer, a grayish lizard-eating snake.

The downside of Brodifacoum, from a rodent's perspective, is pretty obvious: internal bleeding is a slow and painful way to go. From an ecologist's perspective, too, there are drawbacks. Non-target animals often take the bait or eat rodents that have eaten it. In this way, poison spreads up and down the food chain. And if just one pregnant mouse survives an application, she can re-populate an entire island.

Gene-drive mice would scuttle around these problems. Impacts would be targeted. There would be no more bleeding to death. And, perhaps best of all, gene-drive rodents could be re-leased on inhabited islands, where dropping anticoagulants from the air is, understandably, frowned upon.

But as is so often the case, solving one set of problems introduces new ones. In this case, big ones. Humongous ones. Gene-drive technology has been compared to Kurt Vonnegut's *ice-nine*, a single shard of which is enough to freeze all the water in the world. A single X-shredder mouse on the loose could, it's feared, have a similarly chilling effect—a sort of *mice-nine*.

To guard against a Vonnegutian catastrophe, various fail-safe schemes have been proposed, with names like "killer-rescue," "multi-locus assortment," and "daisy-chain." All of them share a basic, hopeful premise: that it should be possible to engineer a gene drive that's effective and at the same time not *too* effective. Such a drive might be engineered so as to exhaust itself after a few generations, or it might be yoked to a gene variant that's limited to a single population on a single island. It has also been suggested that if a gene drive did somehow manage to go rogue, it might be possible to send out into the world another gene drive, featuring a so-called CATCHA sequence, to chase it down. What could possibly go wrong?

While I was in Australia, I wanted to get out of the lab and into the countryside. I thought it would be fun to see some northern quolls; in the photos I'd found online, they looked awfully cute—a bit like miniature badgers. But when I asked around, I learned that quoll-spotting required a lot more expertise and time than I had. It would be much easier to find some of the amphibians that were killing them. So one evening I set out with a biologist named Lin Schwarzkopf to go toad hunting.

As it happened, Schwarzkopf was one of the inventors of the Toadinator trap, and we stopped by her office, at James Cook University, to take a look at the device. It was a cage about the size of a toaster oven, with a plastic-flap door. When Schwarzkopf turned on the trap's little speaker, the office reverberated with the toad's thrumming call.

"Male toads are attracted to anything that sounds even remotely like a cane toad," she told me. "If they hear a generator, they'll go to it."

James Cook University is situated on the northern Queensland coast, in the region where the toads were first introduced. Schwarzkopf figured we should be able to locate some toads right on the university grounds. We both strapped on headlamps. It was about 9 P.M., and the campus was deserted, except for the two of us and a family of wallabies hopping about. We wandered around for a while, looking for the glint of a malevolent eye. Just as I was beginning to lose heart, Schwarzkopf spotted a toad in the leaf litter. Picking it up, she immediately identified it as a female.

"They won't hurt you unless you give them a really hard

time," she said, pointing out the toad's toxin glands, which looked like two baggy pouches. "That's why you shouldn't hit them with a golf club. Because if you hit the glands, the poison can spray out. And if it gets in your eyes, it will blind you for a few days."

We wandered around some more. It had been so dry, Schwarzkopf observed, the toads were probably short on moisture: "They love air-conditioning units—anything that's dripping." Near an old greenhouse, where someone had recently run a hose, we found two more toads. Schwarzkopf flipped over a rotting crate the size and shape of a casket. "The mother lode!" she announced. In about a quarter-inch of scummy water were more cane toads than I could count. Some of the toads were sitting on top of one another. I thought they might try to get away; instead, they sat there, unperturbed.

The strongest argument for gene editing cane toads, house mice, and ship rats is also the simplest: what's the alternative? Rejecting such technologies as unnatural isn't going to bring nature back. The choice is not between what was and what is, but between what is and what will be, which, often enough, is nothing. This is the situation of the Devils Hole pupfish, the Shoshone pupfish, and the Pahrump poolfish, of the northern quoll, the Campbell Island teal, and the Tristan albatross. Stick to a strict interpretation of the natural and these—along with thousands of other species—are goners. The issue, at this point, is not whether we're going to alter nature, but to what end?

"We are as gods and might as well get good at it," Stewart Brand, editor of the *Whole Earth Catalog*, famously wrote in its first issue, published in 1968. Recently, in response to the whole-earth transformation that's under way, Brand has sharpened his statement: "We are as gods and *have to* get good at it." Brand has co-founded a group, Revive & Restore, whose stated mission is to "enhance biodiversity through new techniques of genetic rescue." Among the more fantastic projects the group has backed is an effort to resurrect the passenger pigeon. The idea is to reverse history by rejiggering the genes of the bird's closest living relative, the band-tailed pigeon.

Much closer to realization is an effort to bring back the American chestnut tree. The tree, once common in the eastern United States, was all but wiped out by chestnut blight. (The blight, a fungal pathogen introduced in the early twentieth century, killed off nearly every chestnut in North America—an estimated four billion trees.) Researchers at the SUNY College of Environmental Science and Forestry, in Syracuse, New York, have created a genetically modified chestnut that's immune to blight. The key to this resistance is a gene imported from wheat. Owing to this single borrowed gene, the tree is considered transgenic and subject to federal permitting. As a consequence, the blight-resistant saplings are, for now, confined to greenhouses and fenced-in plots.

As Tizard points out, we're constantly moving genes around the world, usually in the form of entire genomes. This is how chestnut blight arrived in North America in the first place; it was carried in on Asian chestnut trees, imported from Japan. If we can correct for our earlier tragic mistake by shifting just one more gene around, don't we owe it to the American chestnut to do so? The ability to "rewrite the very molecules of life" places us, it could be argued, under an obligation.

Of course, the argument against such intervention is also compelling. The reasoning behind "genetic rescue" is the sort responsible for many a world-altering screwup. (See, for example, Asian carp and cane toads.) The history of biological interventions designed to correct for previous biological interventions reads like Dr. Seuss's *The Cat in the Hat Comes Back*, in which the Cat, after eating cake in the bathtub, is asked to clean up after himself:

> Do you know how he did it?
> WITH MOTHER'S WHITE DRESS!
> Now the tub was all clean,
> But her dress was a mess!

In the 1950s, Hawaii's Department of Agriculture decided to control giant African snails, which had been introduced two decades earlier as garden ornaments, by importing rosy wolfsnails, which are also known as cannibal snails. The cannibal snails mostly left the giant snails alone. Instead, they ate their way through dozens of species of Hawaii's small endemic land snails, producing what E. O. Wilson has called "an extinction avalanche."

Responding to Brand, Wilson has observed, "We are not as gods. We're not yet sentient or intelligent enough to be much of anything."

Paul Kingsnorth, a British writer and activist, has put it this way: "We are as gods, but we have failed to get good at it . . . We are Loki, killing the beautiful for fun. We are Saturn, devouring our children."

Kingsnorth has also observed, "Sometimes doing nothing is better than doing something. Sometimes it is the other way around."

UP IN THE AIR

1

A few years ago, I received a sales pitch, via email, from a company offering a new service to those concerned about their role in wrecking the planet. For a price, the company, called Climeworks, would scrub subscribers' carbon emissions from the air. Then it would inject the CO_2 a half a mile underground, where the gas would harden into rock.

"Why turn CO_2 into stone?" the email asked. Because humanity had already emitted so much carbon "that we have to physically remove it from the atmosphere to keep global warming at safe levels." I immediately signed on, becoming a so-called "pioneer." Every month, the company sent me another email—"your subscription will renew soon and you will continue to turn CO_2 emissions into stone"—before billing my credit card. After a year of this, I decided it was time to visit my emissions, an admittedly reckless move that swelled my emissions still further.

Though Climeworks is based in Switzerland, its air-into-rock operation is situated in southern Iceland. Once I got to Reykjavík, I rented a car and drove east along Route 1, the ring road that circles the country. After about ten minutes, I was clear of the city. After about twenty, I was beyond the suburbs, racing across an ancient lava field.

Iceland is essentially all lava field. It sits atop the Mid-Atlantic Ridge, and, as the Atlantic Ocean widens, it's being pulled in opposite directions. Running diagonally across the country is a seam lined with active volcanoes. I was headed toward a spot near the seam—a three-hundred-megawatt geothermal plant known as the Hellisheiði Power Station. The landscape looked as if it had been paved by giants and then abandoned. There were no trees or bushes, just clumps of grass and moss. Squarish black boulders lay jumbled in heaps.

When I arrived at the gate of the plant, the whole place seemed to be steaming and the air stank of sulfur. Soon a cute little car drove up, painted in bright orange. Out of it climbed Edda Aradóttir, a managing director at Reykjavík Energy, which owns the power station. Aradóttir is blond and bespectacled, with a round face and long hair that she was wearing pinned back. She handed me a hard hat and put one on herself.

As power stations go, geothermal plants are "clean." Instead of burning fossil fuels, they rely on steam or superheated water pumped from underground, which is why they tend to be sited in volcanically active areas. Still, as Aradóttir explained to me, they, too, produce emissions. With the superheated water inevitably come unwanted gases, like hydrogen sulfide (responsible for the stink) and carbon dioxide. Indeed, pre-Anthropocene, volcanoes were the atmosphere's chief source of CO_2.

About a decade ago, Reykjavík Energy came up with a plan to make its clean energy even cleaner. Instead of allowing the carbon dioxide to escape into the air, the Hellisheiði plant would capture the gas and dissolve it in water. Then the mixture—basically, high-pressure club soda—would be injected back underground. Calculations done by Aradóttir and others suggested that deep beneath the surface, the CO_2 would react with the volcanic rock and mineralize.

"We know that rocks, they store CO_2," she told me. "They're actually one of the biggest reservoirs of carbon on earth. The idea is to imitate and accelerate this process to fight global climate change."

Aradóttir opened the gate, and we drove in the little orange car to the back of the power station. It was a breezy day in late spring, and the steam rising from the pipes and cooling towers seemed unable to make up its mind which way to blow. We paused at a large metal-clad outbuilding attached to a structure resembling a rocket launcher. A sign on the building said: STEINRUNNIÐ GRÓÐURHÚSALOFT, which was translated as "greenhouse gas petrified." Aradóttir told me that the rocket launcher was where the power station's CO_2 was separated from other geothermal gases and prepared for injection. We drove on a bit farther and came to what looked like an outsized air conditioner stuck onto a shipping container. A sign on the container said: ÚR LAUSU LOFTI, or "out of thin air."

This, Aradóttir said, was the Climeworks machine that was scrubbing my emissions—really, just a fraction of my emissions—from the atmosphere. The machine, formally known as a direct air capture unit, suddenly started to hum. "Oh, the cycle just started," she said. "Lucky us!

"At the beginning of the cycle, the equipment sucks in air," she went on. "The CO_2 sticks to specific chemicals inside the capture unit. We heat up the chemicals and that releases the CO_2." This CO_2—the Climeworks CO_2—is then added to the club-soda mixture from the power plant as it makes its way to the injection site.

Even without any help, most of the carbon dioxide humans have emitted would eventually turn to stone, via a natural process known as chemical weathering. But "eventually" here means hundreds of thousands of years, and who has time to wait for nature? At Hellisheiði, Aradóttir and her colleagues were speeding up the chemical reactions by several orders of magnitude. A process that would ordinarily take millennia to unfold was being compressed into a matter of months.

Aradóttir had brought along a rock core to show me the end result. The core, which was roughly two feet long and a couple of inches in diameter, was the dark color of the lava fields. But the black rock—basalt—was pocked with little holes, and these holes were filled with a chalky white compound—calcium carbonate. The white deposits represented, if not my own emissions, then at least somebody's.

A basalt core with pockets of calcium carbonate

• • •

When, exactly, people began altering the atmosphere is a matter of debate. According to one theory, the process got under way eight or nine thousand years ago, before the dawn of recorded history, when wheat was domesticated in the Middle East and rice in Asia. Early farmers set to clearing land for agriculture, and as they chopped and burned their way through the forests, carbon dioxide was released. The quantities involved were comparatively small, but, according to advocates of this theory, known as the "early Anthropocene hypothesis," the effect was fortuitous. Owing to natural cycles, CO_2 levels should have been falling during this period. Human intervention kept them more or less constant.

"The start of the switchover from control of climate by nature to control by humans occurred several thousand years ago," William Ruddiman, a professor emeritus at the University of Virginia and the most prominent proponent of an "early Anthropocene," has written.

According to a second, more widely held view, the switchover only really started in the late-eighteenth century, after the Scottish engineer James Watt designed a new kind of steam engine. Watt's engine, it's often said, anachronistically, "kick-started" the Industrial Revolution. As water power gave way to steam power, CO_2 emissions began to rise, at first slowly, then vertiginously. In 1776, the first year Watt marketed his invention, humans emitted some fifteen million tons of CO_2. By 1800, that figure had risen to thirty million tons. By 1850 it had increased to two hundred million tons a year and by 1900 to almost two billion. Now, the figure is close to forty billion tons annually. So much have we altered the atmosphere that one out of every three molecules of CO_2 loose in the air today was put there by people.

Thanks to this intervention, average global temperatures have, since Watt's day, risen by 1.1° Celsius (2° Fahrenheit). This has led to a variety of increasingly unhappy consequences. Droughts are growing deeper, storms fiercer, heat waves deadlier. Wildfire season is getting longer and the fires more intense. The rate of sea-level rise is accelerating. A recent study in the journal *Nature* reported that, since the 1990s, melt off of Antarctica has increased threefold. Another recent study predicted that most atolls will, in another few decades, become uninhabitable; this includes entire nations, like the Maldives and the Marshall Islands. To paraphrase J. R. McNeill paraphrasing Marx, "Men make their own climate, but they do not make it just as they please."

No one can say exactly how hot the world can get before out-and-out disaster—the inundation of a populous country like Bangladesh, say, or the collapse of crucial ecosystems like coral reefs—becomes inevitable. Officially, the threshold of catastrophe is an average global temperature rise of 2°C (3.6°F). Virtually every nation signed on to this figure at a round of climate negotiations held in Cancún in 2010.

Meeting in Paris in 2015, world leaders had second thoughts. The two-degree threshold, they decided, was too high. The signatories of the Paris Agreement committed themselves to "holding the increase in the global average temperature to well below 2°C . . . and pursuing efforts to limit the temperature increase to 1.5°C."

In either case, the math is punishing. To stay under 2°C, global emissions would have to fall nearly to zero within the next several decades. To stave off 1.5°C, they'd have to drop most of the way toward zero within a single decade. This would entail, for starters: revamping agricultural systems, transforming manufacturing, scrapping gasoline- and diesel-powered vehicles, and replacing most of the world's power plants.

Carbon dioxide removal offers a way to change the math. Extract large amounts of CO_2 from the atmosphere and "negative emissions" could, conceivably, balance out the positive variety. It might even be feasible to cross the threshold of catastrophe and then suck enough carbon out of the air to keep calamity at bay, a situation that's become known as "overshoot."

If anyone can be said to have invented "negative emissions," it's a German-born physicist named Klaus Lackner. Lackner, who's now in his late sixties, is a trim man with dark eyes and a prominent forehead. He works at Arizona State University, in Tempe, and I met up with him one day at his office there. The office was almost entirely bare, except for a few *New Yorker* cartoons on the theme of nerd-dom, which, Lackner told me, his wife had cut out for him. In one of the cartoons, a couple of scientists stand in front of an enormous whiteboard covered in equations. "The math is right," the first scientist says. "It's just in poor taste."

Lackner has lived in the United States for most of his adult life. In the late 1970s, he moved to Pasadena to study with George Zweig, one of the discoverers of quarks, and a few years later, he moved to the Los Alamos National Laboratory, to do research on fusion. "Some of the work was classified," he told me, "some of it not."

Fusion is the process that powers the stars and, closer to home, thermonuclear bombs. When Lackner was at Los Alamos, it was being touted as the energy source of the future. A fusion reactor could generate essentially limitless quantities of carbon-free power from isotopes of hydrogen. Lackner became convinced that a fusion reactor was, at a minimum, decades away. Decades later, it's generally agreed that a workable reactor is still decades away.

"I realized, probably earlier than most, that the claims of the demise of fossil fuels were greatly exaggerated," Lackner told me.

One evening in the early 1990s, Lackner was having a beer with a friend, Christopher Wendt, who's also a physicist. The two got to wondering why, as Lackner put it to me, "nobody's doing these really crazy, big things anymore." This led to more questions and more conversations (and possibly also more beers).

The two came up with their own "crazy, big" idea, which, they decided, wasn't really so crazy. A few years after the original conversation, they produced an equation-dense paper in which they argued that self-replicating machines could satisfy the world's energy needs, and, more or less at the same time, clean up the mess humans had created by burning fossil fuels. They called the machines "auxons," from the Greek *αυξάνω*, meaning "grow." The auxons would be powered by solar panels and, as they multiplied, they'd produce more solar panels, which they'd assemble using elements, like silicon and aluminum, extracted from ordinary dirt. The expanding collection of panels would produce ever more power, at a rate that would increase exponentially. An array covering three hundred eighty-six thousand square miles, an area as large as Nigeria but, as Lackner and Wendt noted, "smaller than many deserts," could meet all the globe's electricity demands many times over.

This same array could also be put to use scrubbing carbon. A Nigeria-sized solar farm would, they calculated, be sufficient to remove all the carbon dioxide emitted by humans up to that point. Ideally, the CO_2 would be converted to rock, much the same way my emissions had been converted in Iceland. Only instead of little pockets of calcium carbonate, there'd be whole countries' worth of it—enough to cover Venezuela in a layer a foot and a half deep. (Where this rock would go, the two did not specify.)

More years went by. Lackner let the auxon idea slide. But he found himself more and more interested in negative emissions.

"Sometimes by thinking through this extreme endpoint you learn a lot," he told me. He began giving talks and writing papers on the subject. Humanity, he said, was going to have to find a way to pull carbon out of the air. Some of his fellow scientists decided he was nuts, others that he was a visionary. "Klaus is, in fact, a genius," Julio Friedmann, a former deputy energy secretary who now works at Columbia University, told me.

In the mid-2000s, Lackner pitched a plan for developing a carbon-sucking technology to Gary Comer, a founder of Lands' End. Comer brought to the meeting his investment adviser, who quipped that what Lackner was looking for wasn't so much venture capital as "adventure capital." Nevertheless, Comer put up $5 million. The company got as far as building a small prototype, but just as it was looking for new investors, the financial crisis of 2008 hit.

"Our timing was exquisite," Lackner told me. Unable to raise more funds, the company folded. Meanwhile, fossil-fuel consumption continued to rise, and along with it, CO_2 levels. Lackner came to believe that, unwittingly, humanity had already committed itself to carbon dioxide removal.

"I think that we're in a very uncomfortable situation," he told me. "I would argue that if technologies to pull CO_2 out of the environment fail, then we're in deep trouble."

Lackner founded the Center for Negative Carbon Emissions at ASU in 2014. Most of the equipment he dreams up is put together in a workshop a few blocks from his office. After we had chatted for a while, we walked over there.

In the workshop, an engineer was tinkering with what looked like the guts of a foldout couch. Where, in the living-room version, there would have been a mattress, in this one was an elaborate array of plastic ribbons. Embedded in each ribbon was a powder made from thousands upon thousands of tiny amber-colored beads. The beads, Lackner explained, were composed of a resin normally used in water treatment and could be purchased by the truckload. When dry, the powder made from the beads would absorb carbon dioxide. When wet, it would release it. The idea behind the couch-like arrangement was to expose the ribbons to Arizona's thirsty air, then fold the device into a sealed container filled with water. The CO_2 that had been captured in the dry phase would be released in the wet phase; it could then be piped out of the container and the whole process restarted, the couch folding and unfolding over and over again.

Lackner told me he'd calculated that an apparatus the size of a semi-trailer could remove a ton of carbon dioxide per day, or three hundred and sixty-five tons a year. Since global emissions are now running around forty billion tons a year, he observed, "if you built a hundred million trailer-sized units," you could more or less keep up. He acknowledged the hundred-million figure sounded daunting. But, he noted, the iPhone has only been around since 2007, and there are now almost a billion in use. "We are still very early in this game," he said.

The way Lackner sees things, the key to avoiding "deep trouble" is thinking differently. "We need to change the paradigm," he told me. Carbon dioxide, in his view, should be regarded much the same way we look at sewage. We don't expect people to stop producing waste. "Rewarding people for going to the bathroom less would be nonsensical," Lackner has observed. At the same time, we don't let them shit on the sidewalk. One of the reasons we've had such trouble addressing the carbon problem, he contends, is the issue has acquired an ethical charge. To the extent that emissions are seen as bad, emitters become guilty.

"Such a moral stance makes virtually everyone a sinner and makes hypocrites out of many who are concerned about climate change but still partake in the benefits of modernity," he has written. Shifting the paradigm, he thinks, will shift the conversation. Yes, people have fundamentally altered the atmosphere. And, yes, this is likely to lead to all sorts of dreadful consequences. But people are ingenious. They come up with crazy, big ideas, and sometimes these actually work.

During the first few months of 2020, a vast, unsupervised experiment took place. As the coronavirus raged, billions of people were ordered to stay home. At the peak of the lockdown, in April, global CO_2 emissions were down an estimated seventeen percent compared with the comparable period the previous year.

This drop—the largest recorded ever—was immediately followed by a new high. In May 2020, carbon dioxide levels in the atmosphere set a record of 417.1 parts per million.

Declining emissions and rising atmospheric concentrations point to a stubborn fact about carbon dioxide: once it's in the air, it stays there. How long, exactly, is a complicated question; for all intents and purposes, though, CO_2 emissions are cumulative. The comparison that's often made is to a bathtub. So long as the tap is running, a stoppered tub will continue to fill. Turn the tap down, and the tub will still keep filling, just more slowly.

To extend the analogy, it could be said that the 2°C tub is approaching capacity and that the tub for 1.5°C is all-but-overflowing. This is why the carbon math is so difficult. Cutting emissions is at once absolutely essential and insufficient. Were we to halve emissions—a step that would entail rebuilding much of the world's infrastructure—CO_2 levels wouldn't drop; they'd simply rise less quickly.

Then there's the issue of equity. Since carbon emissions are cumulative, those most to blame for climate change are those who've emitted the most over time. With just four percent of the world's population, the United States is responsible for almost thirty percent of aggregate emissions. The countries of the European Union, with about seven percent of the globe's population, have produced about twenty-two percent of aggregate emissions. For China, home to roughly eighteen percent of the globe's population, the figure is thirteen percent. India, which is expected soon to overtake China as the world's most populous nation, is responsible for about three percent. All the nations of Africa and all the nations of South America put together are responsible for less than six percent.

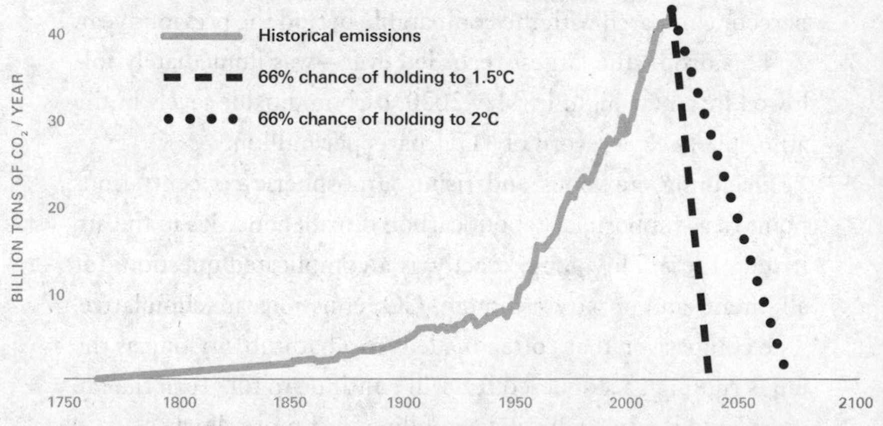

For the world to have a two-thirds chance of staying under 2°C without carbon dioxide removal, CO_2 emissions would have to fall to zero within the next several decades. To stay under 1.5°C, emissions would have to fall much faster.

To get to zero, everyone would have to stop emitting—not only Americans and Europeans and Chinese, but also Indians and Africans and South Americans. But asking countries that have contributed almost nothing to the problem to swear off carbon because other countries have already produced way, way too much of it is grossly unfair. It's also geopolitically untenable. For this reason, international climate agreements have always been based on the premise of "common but differentiated responsibilities." Under the Paris accord, developed countries are supposed to "lead by undertaking economy-wide absolute emission reduction targets," while developing countries are called on, more hazily, to enhance their "mitigation efforts."

All of which makes negative emissions—as an idea at least—irresistible. The extent to which humanity is already counting on them is illustrated by the latest report of the Intergovernmental Panel on Climate Change, which was published in the run-up to Paris. To peer into the future, the IPCC relies on computer models that represent the world's economic and energy systems as a tangle of equations. The output of these models is then translated into figures that climate scientists can use to forecast how much temperatures are going to rise. For its report, the IPCC considered more than a thousand scenarios. The majority of these led to temperature increases beyond the official 2°C disaster threshold, and some led to warming of more than 5°C (9° Fahrenheit). Just a hundred and sixteen scenarios were consistent with holding warming under 2°C, and of these, a hundred and one involved negative emissions. Following Paris, the IPCC produced another report, based on the 1.5°C threshold. *All* of the scenarios consistent with that goal relied on negative emissions.

"I think what the IPCC really is saying is, 'We tried lots and lots of scenarios,'" Klaus Lackner told me. "'And, of the scenarios which stayed safe, virtually every one needed some magic touch of negative emissions. If we didn't do that, we ran into a brick wall.' "

Climeworks, the company I paid to bury my emissions in Iceland, was founded by two college friends, Christoph Gebald and Jan Wurzbacher. "We met on the first day of starting university," Wurzbacher recalled. "I think we asked each other in the first week, 'Hey, what do you want to do?' And I said, 'Well, I want to found my own company.' " The pair ended up splitting a single graduate school stipend; each worked half-time on his PhD and half-time on getting their company off the ground.

Like Lackner, the two initially faced a lot of skepticism. What

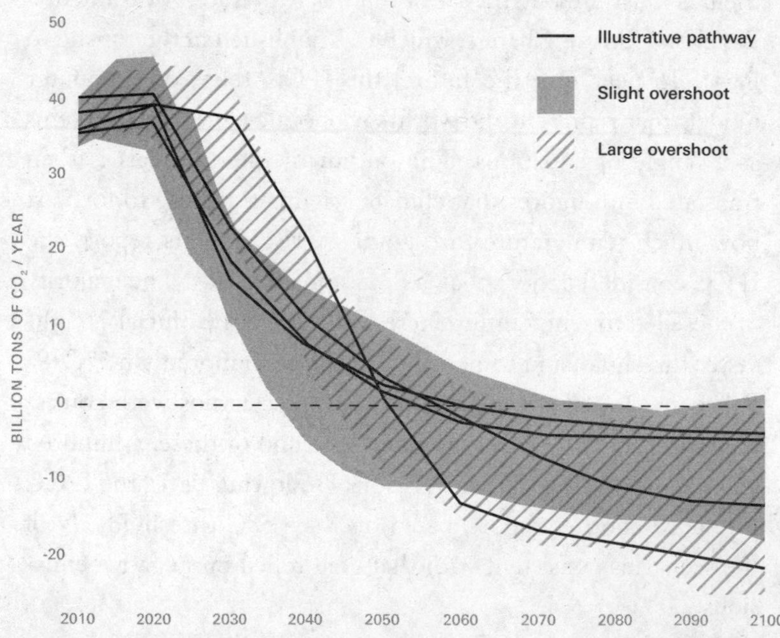

Four of the IPCC's "illustrative 1.5° C-consistent" pathways. All of the pathways require negative emissions and result in "overshoot."

the duo was trying to do, they were told, was a distraction. If people thought there was some way to draw carbon dioxide out of the atmosphere, they'd just emit even more of it. "People were fighting us, saying, 'Well, guys, you shouldn't be doing that,'" Wurzbacher told me. "But we were always stubborn."

Wurzbacher, who's now in his mid-thirties, is reed-thin, with a boyish mop of dark hair. I met with him at Climeworks' headquarters in Zurich, which houses both the company's offices and its metalworking shop. The place had some of the vibe of a tech start-up and some of the vibe of a bike store.

"Taking CO_2 out of a gas stream, that's not rocket science," Wurzbacher told me. "And it's also not new. People have filtered CO_2 out of gas streams for the last fifty years, just for other applications." On submarines, for instance, the carbon dioxide the crew breathes out has to be sucked from the air; otherwise, it will build up to dangerous levels.

But it's one thing to be able to pull carbon out of the air and quite another to be able to pull this off at scale. Burning fossil fuels generates energy. Capturing CO_2 from the air *requires* energy. So long as this energy comes from burning fossil fuels, it will add to the carbon that has to be captured.

A second major challenge is disposal. Once captured, CO_2 has to go somewhere, and where it goes has to be secure. "The good thing about basaltic rock is it's so easy to explain," Wurzbacher observed. "If someone asks you, 'Hey, but is it really safe?' the answer is very simple: within two years it's stone, one kilometer underground. Period." Suitable underground storage sites aren't rare, but they aren't common, either, meaning that, should large-scale capture plants ever be built, they'll either have to be located in places with the right geology or the CO_2 will have to be shipped long distances.

Finally, there's the issue of cost. Pulling CO_2 from the air takes money. Right now, a lot of money. Climeworks charges $1,000 a ton to turn subscribers' emissions to stone. I used up my allotment of twelve hundred pounds to fly one-way to Reykjavík, leaving all the rest of my emissions, including those from my return trip and my flight to Switzerland, floating loose. Wurzbacher assured me that, as more capture units went up, the price would come down; within a decade or so, he predicted, it would fall to around $100 per ton. Were emissions taxed at a comparable rate, then the math could work out: Basically, a ton extracted would be a ton that could avoid the tax. But who's going to spend that when carbon can still be dumped in the air for free? Even at $100 a ton, burying a billion tons of CO_2—a small percentage of the world's annual output—would run to $100 billion.*

"Maybe we are too early," Wurzbacher mused, when I asked whether the world was prepared to pay for direct air capture. "Maybe we're just right. Maybe we're too late. No one knows."

Just as there are lots of ways to add CO_2 to the air, there are lots of ways—potentially—to remove it.

A technique known as "enhanced weathering" is a sort of upside-down version of the project I toured at the Hellisheiði

* There are two ways to measure quantities of CO_2: by accounting for either the full weight of the carbon dioxide or just the weight of the carbon. In this chapter, I am generally using the former measure, as is Climeworks, but many scientific publications use the latter. I have tried to distinguish the two by referring to a "ton of carbon dioxide" when I mean the full weight, and a "ton of carbon" when I mean the alternative. One ton of carbon dioxide translates into roughly a quarter of a ton of carbon; thus, annual global emissions are either about forty billion tons of CO_2 or ten billion tons of carbon.

Power Station. Instead of injecting CO_2 deep into rock, the idea is to bring the rock up to the surface to meet the CO_2. Basalt could be mined, crushed, and then spread over croplands in hot, humid parts of the world. The crushed stone would react with carbon dioxide, drawing it out of the air. Alternatively, it's been proposed that olivine, a greenish mineral that's common in volcanic rock, could be ground up and dissolved in the oceans. This would induce the seas to absorb more CO_2 and, as an added benefit, combat ocean acidification.

Another family of negative-emissions technologies, or NETs, takes its cue from biology. Plants absorb carbon dioxide while they're growing; then, when they rot, they return that CO_2 to the air. Grow a new forest and it will draw down carbon until it reaches maturity. A recent study by Swiss researchers estimated that planting a trillion trees could remove two hundred billion tons of carbon from the atmosphere over the next several decades. Other researchers argued that this figure overstated the case by a factor of ten or even more. Nevertheless, they observed, the capacity of new forests to sequester carbon was "still substantial."

To deal with the rot problem, all sorts of preservation techniques have been proposed. One entails cutting down mature trees and burying them in trenches; in the absence of oxygen, the trees' decay—and the release of CO_2—would be forestalled. Another scheme involves collecting crop residues, like cornstalks, and dumping them deep into the ocean; in the dark, cold depths, the waste material would decay very gradually or perhaps not at all. As strange as these ideas may sound, they, too, take their inspiration from nature. In the Carboniferous period, vast quantities of plant material got flooded and buried. The eventual result was coal, which, had it been left in the ground, would have held on to its carbon more or less forever.

Reforestation, when combined with underground injection, yields a technique that's become known as BECCS (pronounced "becks"), short for "bioenergy with carbon capture and storage." The models employed by the IPCC are extremely partial to BECCS, which offers negative emissions and electrical power at the same time—a have-your-cake-and-eat-it-too arrangement that, in climate-math terms, is tough to beat.

With BECCS the idea is to plant trees (or some other crop) that can pull carbon from the air. The trees are then burned to produce electricity and the resulting CO_2 is captured from the smokestack and shoved underground. (The world's first BECCS pilot project launched in 2019, at a power plant in northern England that runs off wood pellets.)

With all of these alternatives, the challenge is much the same as with direct air capture: scale. Ning Zeng is a professor at the University of Maryland and the author of the "wood harvest and storage" concept. He has calculated that to sequester five billion tons of carbon per year, ten million tree-burial trenches, each the size of an Olympic swimming pool, would be required. "Assuming it takes a crew of ten people (with machinery) one week to dig a trench," he has written, "two hundred thousand crews (two million workers) and sets of machinery would be needed."

According to a recent study by a team of German scientists, to remove a billion tons of CO_2 through "enhanced weathering," approximately three billion tons of basalt would have to be mined, crushed, and transported. "While this is a very large amount" of rock to mine, grind, and ship, the authors noted, it is less than global coal production, which totals some eight billion tons per year.

For the trillion-tree project, something on the order of 3.5 million square miles of new forest would be needed. That's an expanse of woods roughly the size of the United States, including Alaska. Take that much arable land out of production and millions could be pushed toward starvation. As Olúfẹ́mi O. Táíwò, a professor at Georgetown, put it recently, there's a danger of moving "two steps backward in justice for every gigaton step forward." But it's not clear that using uncultivated land would be any safer. Trees are dark, so if, say, tundra were converted to forest, it would increase the amount of energy being absorbed by the earth, thus contributing to global warming and defeating the purpose. One way around this problem might be to genetically engineer lighter-colored trees, using CRISPR. So far as I know, no one has yet proposed this, but it seems only a matter of time.

A couple of years before Climeworks launched its "pioneer" program in Iceland, the company opened its first direct-air-capture operation, atop a garbage incinerator in Switzerland. "Climeworks makes history," the company declared.

One afternoon while I was in Zurich, I went to visit the "history-making" operation with Climeworks' communications manager, Louise Charles. We took a train and then a bus out to the town of Hinwil, about twenty miles southeast of the city. As we walked up the access road to the incinerator, a huge box of a building with a candy-striped smokestack, a truck rolled by filled with rubbish. In the entrance hall, we paused to admire a series of artworks, also made of rubbish. Several men were seated before video monitors that displayed more rubbish. We signed the visitors' log and took a service elevator up to the top floor.

On the roof of the incinerator were eighteen capture units just like the one at the Hellisheiði plant. These were arranged in three rows, which were stacked one above the other, like children's blocks. A metal placard, aimed at visiting school groups,

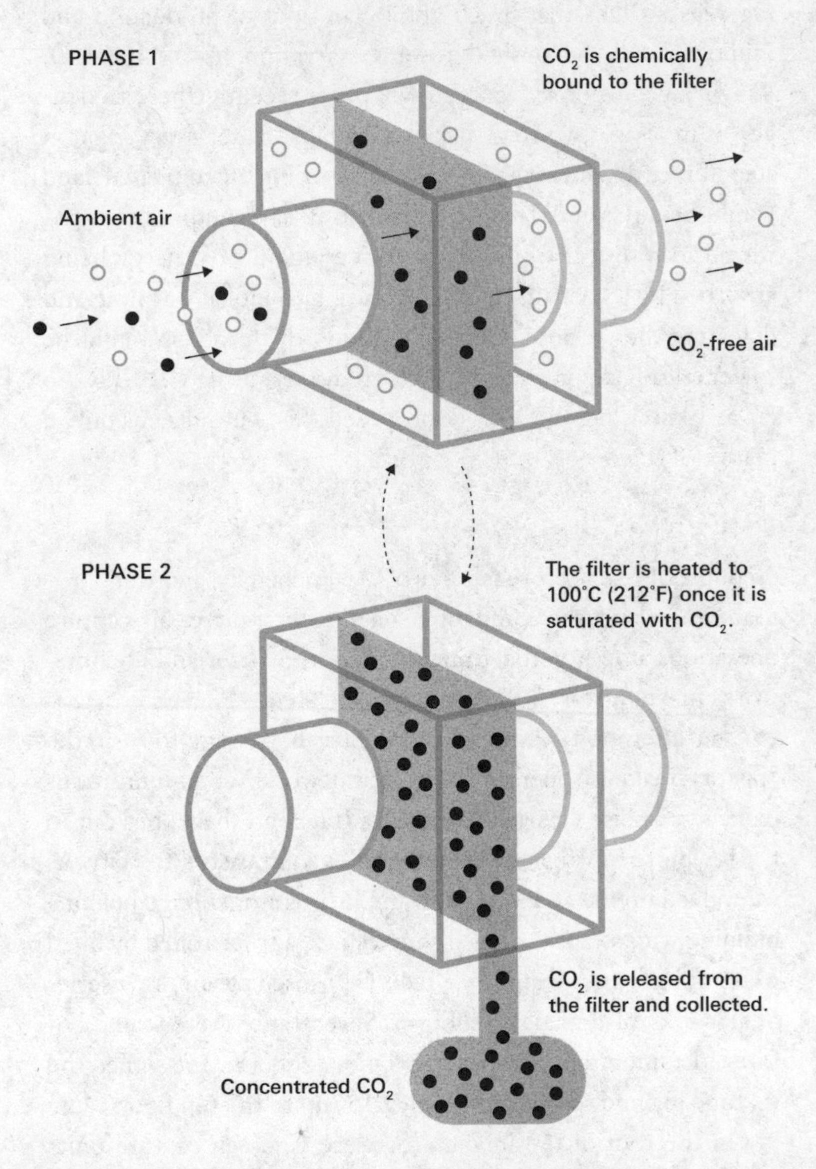

Climeworks' carbon dioxide removal system uses a two-step process.

explained the Climeworks operation in pictures. It showed a garbage truck pulling up to the incinerator, which was depicted with little flames inside. One pipe, labeled WASTE HEAT, led from the flames to the stack of capture units. (Using waste heat from the incinerator allows Climeworks to sidestep the it-takes-emissions-to-catch-emissions trap.) A second pipe, labeled CONCENTRATED CO_2, led from the units to a greenhouse filled with floating vegetables.

From the roof, I could see in the distance the actual greenhouses where the CO_2 was headed. Charles had arranged for us to tour those as well, but she'd recently had knee surgery and was hobbling painfully, so I walked over alone. I was met at the entrance by the manager of the complex, Paul Ruser. Without Charles to translate, we had to make do with a hodgepodge of English and German.

Ruser told me—or at least I think he told me—that the greenhouses covered an area of eleven acres: an entire farm, under glass. Outside, it was sweater weather; inside, it was summertime. Bumblebees, which had been imported in boxes, buzzed around groggily. Twelve-foot-tall cucumber vines rose out of small bricks of potting soil. The cucumbers—a miniature variety the Swiss call *Snack-Gurken*—had just been picked and were piled high in bins. Ruser pointed out a black plastic tube running along the floor. This, he explained, was carrying CO_2 from the Climeworks units.

"All plants need CO_2," Ruser observed. "And if you supply more to them, they become stronger." Eggplants in particular, he said, thrive on lots of carbon dioxide; for their sake, he might crank the level way up, to as much as a thousand parts per million—more than double the level in the outside world. He needed to be careful, though. He was paying Climeworks for the piped-in CO_2, so he had to make every molecule count: "I have to figure out the level that's going to be profitable."

Carbon dioxide removal may be essential; it's already built into the calculations of the IPCC. Under the current order, however, it's also economically infeasible. How do you go about creating a $100 billion industry for a product no one wants to buy? The eggplants and the *Snack-Gurken* represented an admittedly jury-rigged solution. By selling its CO_2 to the greenhouses, Climeworks had secured a revenue stream to underwrite its capture units. The catch was that the captured carbon was only briefly being captured. Whoever snacked on the *Snack-Gurken* would liberate the CO_2 that had gone into producing them.

From more little bricks of dirt, cherry tomato plants stretched to the roof in helical coils. The tomatoes, just a day or two from harvest, were perfect, in that greenhouse tomato-y sort of way. Ruser picked a couple and handed them to me. The burning trash, the acres of glass, the boxes of bumblebees, the vegetables raised on chemicals and captured CO_2—was it all totally cool or totally crazy? I paused for a second, then popped the tomatoes into my mouth.

2

The Volcanic Explosivity Index was developed in the 1980s as sort of a cousin to the Richter scale. The index runs from zero, for a gentle burp of an eruption, to eight, for a "mega-colossal," epoch-making catastrophe. Like its better-known relative, the VEI is logarithmic, so, for example, an eruption has a magnitude of four if it produces more than a hundred million cubic meters of ejecta and a magnitude of five if it produces more than a billion. In recorded history, there have been only a handful of magnitude sevens (a hundred billion cubic meters) and no eruptions of magnitude eight. Among the sevens, the most recent—and, hence, the best chronicled—is the eruption of Mount Tambora, on the Indonesian island of Sumbawa.

Tambora fired its first warning shots on the evening of April 5, 1815. People across the region reported hearing loud booms, which they attributed to cannon fire. Five days later, the mountain issued a column of smoke and lava that reached a height of twenty-five miles. Ten thousand people were killed more or less immediately—burned to cinders by the clouds of molten rock and searing vapor that raced down the slopes. One survivor reported seeing "a body of liquid fire, extending itself in every direction." So much dust was thrown into the air that, it's said, day turned to night. According to a British sea captain whose ship was anchored two hundred and fifty miles to the north of Tambora, "It was impossible to see your hand when held up close to the eye." Crops on Sumbawa and the neighboring island of Lombok were buried under ash, leaving tens of thousands more to perish from starvation.

The eruption of Mount Tambora left an enormous crater.

These deaths were just the beginning. Along with ash, Tambora released more than a hundred million tons of gas and fine particles, which remained suspended in the atmosphere for years, drifting around the world on stratospheric winds. The haze itself was invisible; its results were just the opposite. Sunsets in Europe glowed eerily in blue and red, an effect recorded in private diaries and in the works of painters like Caspar David Friedrich and J.M.W. Turner.

Europe's weather turned gray and cold. In what is probably the world's most famous summer share, Lord Byron rented a villa on Lake Geneva in June 1816, with Percy and Mary Shelley as his housemates. Confined indoors by the season's ceaseless rain, they decided to write ghost stories, an exercise that gave birth to *Frankenstein*. That same summer, Byron composed his poem "Darkness," which runs, in part:

> Morn came and went—and came, and brought no day,
> And men forgot their passions in the dread
> Of this their desolation; and all hearts
> Were chill'd into a selfish prayer for light.

The grim weather caused harvests to fail from Ireland to Italy. Traveling through the Rhineland, the military tactician Carl von Clausewitz saw "ruined figures, scarcely resembling men, prowling around the fields," searching for something edible among the "half-rotten potatoes." In Switzerland, hungry crowds destroyed bakeries; in England, protesters marching under the banner BREAD OR BLOOD clashed with police.

How many people starved to death is unclear; some estimates put the figure in the millions. Hunger prompted many Europeans to immigrate to the United States, but conditions on the other side of the Atlantic, it turned out, weren't much better. In New England, 1816 became known as the "year without a summer" or "eighteen-hundred-and-froze-to-death." In mid-June it was so cold in central Vermont that foot-long icicles dripped from the eaves. "The very face of nature," opined the *Vermont Mirror*, "appears to be shrouded in a death-like gloom." On July 8, there was frost as far south as Richmond, Virginia. Chester Dewey, a professor at Williams College, in Williamstown, Massachusetts, where I happen to live, recorded a freeze on August 22 that killed the cucumber crop. A harder freeze on August 29 killed most of the corn.

"What a volcano does is put sulfur dioxide into the stratosphere," Frank Keutsch said. "And that gets oxidized on the scale of weeks to sulfuric acid.

"Sulfuric acid," he continued, "is a very sticky molecule. And it starts making particulate matter—concentrated sulfuric acid droplets—usually smaller than one micron. These aerosols stay in the stratosphere on the timescale of a few years. And they scatter sunlight back to space." The result is lower temperatures, fantastic sunsets, and, on occasion, famine.

Keutsch is a burly man with floppy black hair and a lilting German accent. (He grew up near Stuttgart.) On a lovely late-winter day I went to visit him in his office in Cambridge, which is decorated with pictures of and by his kids. A chemist by training, Keutsch is one of the leading scientists with Harvard's Solar Geoengineering Research Program, an effort funded, in part, by Bill Gates.

The premise behind solar geoengineering—or, as it's sometimes more soothingly called, "solar radiation management"—is that if volcanoes can cool the world, people can, too. Throw a gazillion reflective particles into the stratosphere and less sunlight will reach the planet. Temperatures will stop rising—or at least not rise as much—and disaster will be averted.

Even in an age of electrified rivers and redesigned rodents, solar geoengineering is out there. It has been described as "dangerous beyond belief," "a broad highway to hell," "unimaginably drastic," and also as "inevitable."

"I thought the idea was entirely crazy and quite disconcerting," Keutsch told me. What brought him around was fear.

"The thing I worry about is that in ten or fifteen years, people could go out in the street and demand from decision-makers, 'You guys need to take action now!'" he said. "We have this integrated CO_2 problem that you can't do anything about very quickly. So if there's pressure from the public to do something fast, my concern is that there will be no tools at hand other than stratospheric geoengineering. And if we start doing research at that point, I am concerned it's too late, because with stratospheric geoengineering, you're interfering with a highly complex system. I will add that there are a number of people who do not agree with this.

"When I started this, I was perhaps, oddly, not as worried about it," he observed a few minutes later. "Because the idea that geoengineering would actually happen seemed quite remote. But, over the years, as I see our lack of action on climate, I sometimes get quite anxious that this may actually happen. And I feel quite a lot of pressure from that."

The stratosphere might be thought of as earth's second balcony. It sits above the troposphere, which is where clouds billow, trade winds blow, and hurricanes rage, and beneath the mesosphere, which is where meteors go to vaporize. The height of the stratosphere varies according to the season and the location; very roughly speaking, at the equator, the bottom of the strato-

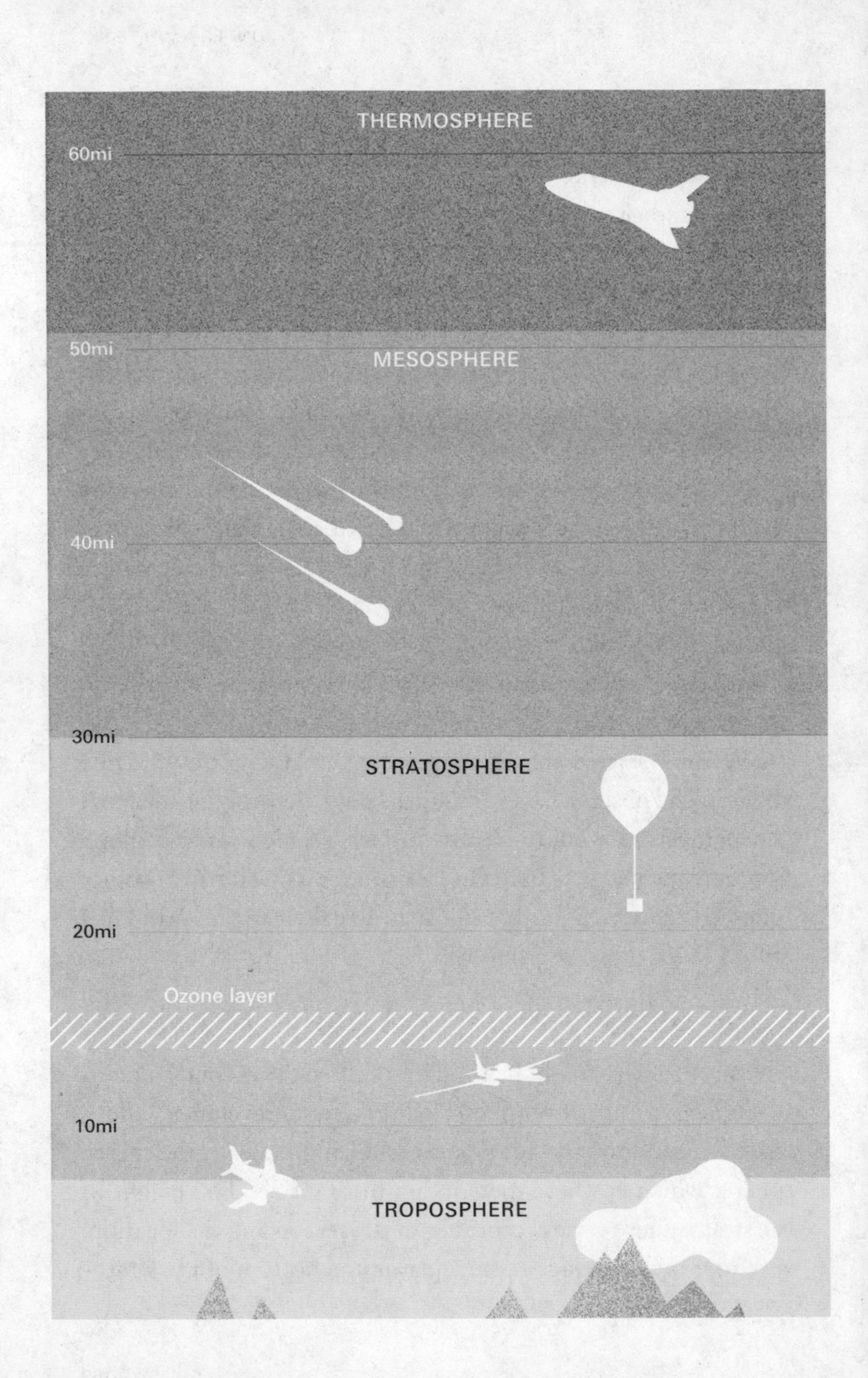
THERMOSPHERE
60mi
50mi
MESOSPHERE
40mi
30mi
STRATOSPHERE
20mi
Ozone layer
10mi
TROPOSPHERE

sphere sits about eleven miles above the surface of the earth, and at the poles it sits much lower—about six miles above the surface. From a geoengineering point of view, what's key about the stratosphere is that it's stable—much more stable than the troposphere—and also reasonably accessible. Commercial jets often fly in the lower stratosphere, to avoid turbulence, and spy planes fly toward the middle, to avoid surface-to-air missiles. Materials injected into the stratosphere in the tropics will tend to drift toward the poles and then, after a few years, drop back to earth.

Since the point of solar geoengineering is to reduce the amount of energy reaching the earth, any sort of reflective particle, in principle at least, would do. "The best possible material probably is diamond," Keutsch told me. "Diamonds really will not absorb any energy. So this would minimize the change in stratospheric dynamics. And diamond itself is extremely unreactive. The idea that this is expensive—I don't care about that. If we had to engineer this on a big scale because it solves a big problem, we would figure out a way to do it." Shooting tiny diamonds into the stratosphere struck me as magical, like sprinkling the world with pixie dust.

"But one of the things to think about is that all the material comes back down," Keutsch continued. "Does that mean that people are inhaling these little diamond particles? It's very likely that the amount would be so small it wouldn't be a problem. But, somehow, I really don't like that idea."

Another option is to go full-on volcano and spray sulfur dioxide. Here, too, there are downsides. Loading the stratosphere with sulfur dioxide would contribute to acid rain. More significantly, it could damage the ozone layer. Following the eruption of Mount Pinatubo, in the Philippines, in 1991, there was a brief downturn in global temperatures of about 1°F. In the tropics, ozone levels in the lower stratosphere fell by as much as a third.

"Perhaps this is not a good phrase, but it's the devil that we know," Keutsch said.

Of all the substances that might be deployed, Keutsch was most enthusiastic about calcium carbonate. In one form or another, calcium carbonate turns up everywhere—in coral reefs, in the pores of basalt, in the ooze at the bottom of the ocean. It's the main component of limestone, which is one of the world's most common sedimentary rocks.

"There are vast amounts of limestone dust blowing around in the troposphere, where we live," Keutsch observed. "So that makes it attractive.

"It has near-ideal optical properties," he went on. "It dissolves in acid. So I can say with certainty that it will not have the same ozone-depleting impact that sulfuric acid has."

Mathematical modeling has confirmed the mineral's advantages, Keutsch told me. But until someone actually throws calcium carbonate into the stratosphere, it's hard to know how much to trust the models. "There's no other way around it," he said.

The first government report on global warming—though the phenomenon was not yet called "global warming"—was delivered to President Lyndon Johnson in 1965. "Man is unwittingly conducting a vast geophysical experiment," it asserted. The result of burning fossil fuels would, almost certainly, be "significant changes in the temperature," which would, in turn, lead to other changes.

"The melting of the Antarctic ice cap would raise sea level by four hundred feet," the report noted. Even if the process took a thousand years to play out, the oceans would "rise about four feet every ten years," or "forty feet per century."

Carbon emissions in the 1960s were growing fast—by about five percent a year. And yet the report made no mention of reversing or even just trying to slow this growth. Instead, it advised that "the possibilities of deliberately bringing about countervailing climatic changes . . . be thoroughly explored." One such possibility was "spreading very small reflecting particles over large oceanic areas.

"Rough estimates indicate that enough particles to cover a square mile could be produced for perhaps one hundred dollars," the report stated. "Thus a one percent change in reflectivity might be brought about for about five hundred million dollars a year"—roughly $4 billion a year in today's money. Considering "the extraordinary economic and human importance of climate, costs of this magnitude do not seem excessive," the report concluded.

None of the authors of the report is still alive, so it's impossible to know why the committee jumped straight to a multimillion-dollar dump of reflective particles. Perhaps it was just the zeitgeist. In the 1960s, climate- and weather-control proposals were all the rage, both in the United States and the USSR. Project Stormfury, a collaboration between the U.S. Navy and the Weather Bureau, targeted hurricanes. These, it was believed, could be weakened by sending aircraft to seed the clouds around the eyewall with silver iodide. Operation Popeye, a secret weather-modification scheme run by the Air Force during the Vietnam War, was supposed to increase rainfall over the Ho Chi Minh Trail, once again by seeding clouds with silver iodide. An astonishing twenty-six hundred seeding sorties were flown by the 54th Weather Reconnaissance Squadron before Popeye was exposed in *The Washington Post* and shut down. (A related program—Operation Commando Lava—involved dumping a mix of chemicals on the trail in an effort to destabilize the soil.) Other climate-modification plans pursued at government expense aimed at reducing lighting strikes and suppressing hail.

The Soviets' schemes were, depending on your perspective, even more farsighted or more off-the-wall. In a book titled *Can Man Change the Climate?* an engineer named Petr Borisov suggested melting the Arctic ice cap with a dam across the Bering Strait. Hundreds of cubic miles' worth of cold water could then, somehow or other, be pumped from the Arctic Ocean into the Bering Sea, which would draw in warmer water from

A rendering of the proposed dam across the Bering Strait

the North Atlantic and, according to Borisov's calculations, produce milder winters not just in the polar regions but also in the mid-latitudes.

"What mankind needs is a war against cold, rather than a 'cold war,'" Borisov declared.

Another Soviet scientist, Mikhail Gorodsky, recommended creating a washer-shaped band of potassium particles around the earth, something like the rings of Saturn. The band would be positioned to reflect sunlight in summer. Gorodsky believed this arrangement would result in much warmer winters in the far north and also lead to a thawing of the world's permafrost, a development that he welcomed. *Man Versus Climate*, a survey of this and other Soviet proposals translated into English by a Moscow-based outfit called Peace Publishers, ended with the declaration:

> New projects for transforming nature will be put forward every year. They will be more magnificent and more exciting, for human imagination, like human knowledge, knows no bounds.

In the 1970s, climate engineering fell out of favor. Once again, it's hard to say exactly why. Public concern about the environment probably had something to do with it, as did the growing scientific consensus that cloud-seeding was a bust. Meanwhile, more and more reports were appearing, in both English and Russian, warning that humans were already modifying the climate, and on a massive scale.

In 1974, Mikhail Budyko, a prominent scientist at the Leningrad Geophysical Observatory, published a book titled *Climatic Changes*. Budyko laid out the dangers posed by rising CO_2 levels but argued that their continued climb was inevitable: The only way to hold down emissions was to cut fossil-fuel use, and no nation was likely to do that.

Following this logic, Budyko arrived at the idea of "artificial volcanoes." Sulfur dioxide might be injected into the stratosphere using planes or "rockets and different types of missiles." Budyko wasn't intent on improving on nature, in the fashion of Project Stormfury or damming the Bering Strait. Rather, he was thinking along more revanchist lines, as in the dictum from *The Leopard:* "If we want everything to remain as it is, everything must change."

"In the near future, climate modification will become necessary in order to maintain current climatic conditions," Budyko wrote.

David Keith, a professor of applied physics at Harvard, has been described as "perhaps the foremost proponent of geoengineering," a characterization that he bristles at. "I'm a proponent of reality," he wrote in a letter to the editor of *The New York Times* in 2015. Keith founded the university's Solar Geoengineering Research Program in 2017, and he regularly receives hate mail. Twice he's gotten death threats worrisome enough to report to the police. His office is just down the hall from Keutsch's, in a building known as the Link.

"Solar geoengineering is not a thing you can study in the abstract," he told me when I went to speak to him a few days after I'd visited Keutsch. "It depends on human choices about how we use it. So whenever anybody makes a statement that solar geoengineering will imperil millions or save the world or whatever, you should always ask, 'What solar geoengineering? Done what way?' "

Keith is tall and angular, with a Lincoln-esque beard. An avid mountaineer, he describes himself as a "tinkerer," a "technophile," and "an oddball environmentalist." He grew up in Canada and for about a decade taught at the University of Calgary. While he was working there, he founded a company, Carbon Engineering, which competes with Climeworks on direct air capture. (Carbon Engineering has a pilot plant in British Columbia that I once visited; it has a spectacular view of Mount Garibaldi, a dormant volcano that rises to a height of nine thousand feet.) Nowadays, he splits his time between Cambridge and Canmore, a town in the Canadian Rockies.

Keith believes that the world will eventually cut its carbon emissions if not all the way down to zero, then close to it. He also believes carbon-removal technologies can eventually be scaled up to take care of the rest. But all this—quite possibly—will not be enough. During the period of "overshoot," a great many people will suffer and changes that are, for all intents and purposes, irreversible may occur, like the demise of the Great Barrier Reef.

The best way forward, he argues, is to do everything: cut emissions, work on carbon removal, *and* look a lot more seriously at geoengineering. On the basis of computer modeling, he's proposed that the safest option would be to put up enough aerosols to cut warming in half, rather than to counteract it entirely—what might be called "semi-engineering."

"If you did not try to restore temperatures to pre-industrial levels, then the evidence from, really, all climate models is that most of the big climate hazards that people know about—extreme precipitation, extreme temperatures, changes in water availability, sea-level rise—are reduced," he told me. This is true, he said, "basically everywhere, in the sense that there are no obvious regions that are made worse off. That result, I think, is really stunning."

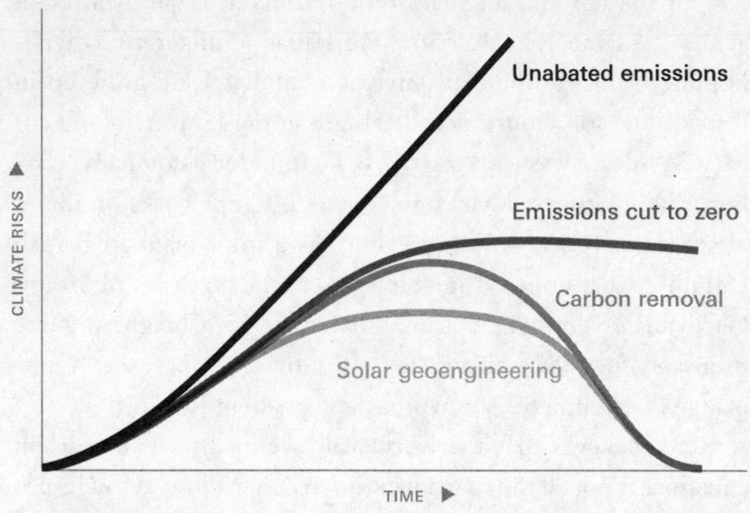

Solar geoengineering could potentially be used to "cut the top off" the risks of climate change.

I asked Keith about what is sometimes called the "moral hazard" problem. If people think geoengineering is going to avert the worst effects of climate change, won't that reduce their motivation to cut emissions? He agreed this was a worry. But he said the opposite was also possible: "opening up the range of options" could inspire *greater* action.

"Moving away from the kind of monomania that says, 'The only thing we can do is cut emissions,' or the more narrow version, which says, 'The only thing we can do is renewables,' I think may actually secure broader political agreement to deal with the problem. People might be *more* willing to spend the big money to cut emissions as part of a project that, overall, wasn't going to just limit the damage but was actually going to make the world better."

I suggested humans didn't have a very good track record when it came to the sort of intervention he was studying. Though importing poisonous amphibians hardly compares to blocking the sun, I cited the example of cane toads.

Keith suggested I was revealing my own biases: "To people who say most of our technological fixes go wrong, I say, 'Okay, did agriculture go wrong?' It's certainly true that agriculture had all sorts of very unexpected outcomes.

"People think of all the bad examples of environmental modification," he went on. "They forget all the ones that are more or less working. There's a weed, tamarisk, originally from Egypt. It's spread all around the desert Southwest and has been very destructive. After a bunch of trials, they imported some bug that eats the tamarisk, and apparently it's kind of working.

"To be clear, I'm not saying that modifications mostly do work. I'm saying it's a wide, undefined set."

Geoengineering is not something you can do with a mail-order kit in your kitchen. Still, as world-altering projects go, it looks to be surprisingly easy. The best method for delivering aerosols would probably be via airplane. The plane would need to be capable of reaching an altitude of around sixty thousand feet and of carrying a payload on the order of twenty tons. Researchers who looked into the configuration of such a craft, which they dubbed a Stratospheric Aerosol Injection Lofter, or SAIL, concluded that development costs would run to about $2.5 billion. This may sound like a lot of money, but it's only about a tenth of what Airbus spent to develop its "superjumbo" A380, a plane it stopped producing after a dozen years. To deploy a fleet of SAILs would cost another $20 billion or so per decade. Again, this is nothing to sneeze at, but the world now spends more than three hundred times that amount every year on fossil-fuel subsidies.

"Dozens of countries would have both the expertise and the money to launch such a program," the researchers—Wake Smith, a lecturer at Yale, and Gernot Wagner, a professor at NYU—observed.

Solar geoengineering would not just be cheap, relatively speaking; it would also be speedy. Pretty much as soon as the fleet of SAILs went into operation, cooling would begin. (A year and a half after Tambora erupted, the cucumbers in New England were frozen.) As Keutsch told me, it's the only way to "do something fast" about climate change.

But if a fleet of SAILs looks like a quick, cut-rate solution, that's primarily because it isn't a solution. What the technology addresses are warming's symptoms, not its cause. For this reason, geoengineering has been compared to treating a heroin habit with methadone, though perhaps a more apt comparison would be to treating a heroin habit with amphetamines. The end result is two addictions in place of one.

Since calcite or sulfate (or diamond) particles lofted into the stratosphere drop back down after a couple of years, they'd need constant replenishing. If the SAILs flew for a few decades and then, for whatever reason—a war, a pandemic, unhappiness with the results—they stopped, the effect would be like opening a globe-sized oven door. All the warming that had been masked would suddenly manifest itself in a rapid and dramatic temperature run-up, a phenomenon that's become known as "termination shock."

Meanwhile, to keep pace with warming, the SAILs would need to deliver bigger and bigger payloads. (In "artificial volcano" terms, this would be the equivalent of staging increasingly violent eruptions.) Smith and Wagner based their cost calculations on the kind of protocol that Keith has proposed, which would halve the rate of warming going forward. The two estimated that around a hundred thousand tons of sulfur would have to be dispersed in the program's first year. By the tenth year, that figure would rise to more than a million tons. During that period, the number of flights would ramp up accordingly, from four thousand a year to more than forty thousand. (Each flight, awkwardly enough, would generate many tons of carbon dioxide, causing more warming, entailing more flights.)

The more particles injected into the stratosphere, the greater the chance of weird side effects. Researchers who looked into using solar geoengineering to offset carbon dioxide levels of five hundred and sixty parts per million—levels that could easily be reached later this century—determined it would change the appearance of the sky. White would become the new blue. The effect, they noted, would cause "the sky over formerly pristine areas to look similar to the sky over urban areas." Another, more felicitous result, they observed, would be glorious sunsets, "similar to those seen after large volcanic eruptions."

Alan Robock is a climate scientist at Rutgers and one of the leaders of the Geoengineering Model Intercomparison Project, or GeoMIP. Robock maintains a list of concerns about geoengineering; the latest version has more than two dozen entries. Number 1 is the possibility that it could disrupt rainfall patterns, causing "drought in Africa and Asia." Number 9 is "less solar electricity generation," and number 17 is "whiter skies." Number 24 is "conflicts between countries." Number 28 is "do humans have the right to do this?"

• • •

For several years, Keith and Keutsch have collaborated on a project known as the Stratospheric Controlled Perturbation Experiment, or SCoPEx (pronounced "scope-ex"). The experiment is supposed to take place somewhere treeless, like the American Southwest, at an altitude of twelve miles. It will feature a pound or two of reflective particles and a zero-pressure balloon attached to a gondola loaded with scientific instruments.

When I visited Cambridge, work on the gondola was under way, and Keith offered to show me the setup. We headed down a maze of halls, into a lab crammed with pipes, gas canisters, packing crates, circuit boards, and a Home Depot's worth of tools. "This is the flight frame," he said, pointing to a shed-sized arrangement of metal beams. "And those are the flight propellers."

Keith explained that the experiment would unfold in stages. First, the unmanned balloon would drift through the stratosphere, releasing a stream of particles from the gondola. Then the balloon would reverse direction and sail back through the plume of particles, so that their behavior could be monitored.

The goal of the experiment is not to test geoengineering per se—a couple of pounds of calcium carbonate or sulfur dioxide is nowhere near enough to make an observable difference to the climate. Nonetheless, SCoPEx would represent the first rigorous field test—or, if you prefer, sky test—of the concept, and there's been a lot of opposition to letting it get off the ground.

"Even if the amount is inconsequential," Keutsch had told me, "it's extremely symbolic to have a balloon in the stratosphere spraying out particles."

"There are people who think that we shouldn't do this experiment for reasons I think are coherent," Keith told me, as we watched one of his graduate students applying epoxy to the landing gear of the SCoPEx gondola. "But the actual physical risk, just to be clear, is that something falls apart and falls on somebody's head."

So far, Harvard's geoengineering research program is the world's best-financed, with funding of almost $20 million. But there are several other research groups in the United States and Europe exploring alternative forms of "climate intervention."

Sir David King, a chemist who served as the chief scientific adviser to British prime ministers Tony Blair and Gordon Brown and as the government's special representative for climate change, recently launched a research initiative, the Centre for Climate Repair, at Cambridge University.

"We're now at about 1.1, 1.2 Celsius above pre-industrial levels," King told me over the phone one day. "And the conclusion is that this is already too much. The Arctic sea ice, for example, has been melting far more rapidly than was predicted. We're seeing the Greenland ice sheet beginning to melt more quickly than was predicted. So how do we cope with this?"

King said that in addition to deep emissions reductions—"without that, frankly, we're cooked"—the center was created to promote research into carbon removal and technologies to "refreeze" the poles. One idea he mentioned was an Arctic version of cloud-brightening. According to this scheme, a fleet of ships would be dispatched to the Arctic Ocean to shoot very fine droplets of salt water into the sky. The salt crystals, it's theorized, would increase the clouds' reflectivity, thus reducing the amount of sunlight striking the ice.

"The hope is to preserve the layer of sea ice that is formed during the polar winter," King said. "And if you proceed with that year on year, you rebuild the ice, layer by layer."

Dan Schrag is the director of the Harvard University Center for the Environment and a MacArthur "genius" grant winner. He helped set up Harvard's geoengineering program and sits on its advisory board.

"Some have expressed consternation at the prospect of engineering the climate for the entire planet," he has written. "Ironically, such engineering efforts may be the best chance for survival for most of the earth's natural ecosystems—although perhaps they should no longer be called natural if such engineering systems are ever deployed."

Schrag's office is about a block away from Keith's and Keutsch's, and while I was visiting Cambridge, I arranged to meet with him there. His dog, Mickey, a genial Chinook, padded over to greet me.

"I don't know if you ever feel pressure like this as a writer," Schrag said. "But I see a lot of pressure from my colleagues to have a happy ending. People want hope. And I'm like, 'You know what? I'm a scientist. My job is not to tell people the good news. My job is to describe the world as accurately as possible.'

"As a geologist, I think about timescales," he went on. "The timescale of the climate system is centuries to tens of thousands of years. If we stop CO_2 emissions tomorrow, which, of course, is impossible, it's still going to warm at least for centuries, because the ocean hasn't equilibrated. That's just basic physics. We're not sure how much additional warming that is, but it could easily be another seventy percent beyond what we've experienced. So in that sense, we're already at 2°C. We're going to be lucky to stop at 4°C. That's not optimistic or pessimistic. I think that's objective reality." (A 4°C global temperature increase—7.2° F—is not just well beyond the official threshold of disaster, it's heading into territory that's probably best described as unthinkable.)

"The idea that somehow research on solar geoengineering is going to open Pandora's box, I think that's just unbelievably naïve," Schrag said. "Do you really believe that the U.S. military or the Chinese military haven't thought about this? Come on! They've done cloud-seeding for rain. This is not a new idea, and it's not a secret.

"People have to get their heads away from thinking about whether they like solar geoengineering or not, whether they think it should be done or not. They have to understand that we don't get to decide. The United States doesn't get to decide. You're a world leader and there's a technology that could take the pain and suffering away, or take some of it away. You've got to be really tempted. I'm not saying they'll do it tomorrow. I feel like we might have thirty years. The highest priority for scientists is to figure out all the different ways this could go wrong."

While we were talking, a friend of Schrag's showed up at his office. Schrag introduced her as Allison Macfarlane, a professor at George Washington University and a former head of the U.S. Nuclear Regulatory Commission. When he told her we were discussing geoengineering, she made a thumbs-down gesture.

"It's the unintended consequences," she said. "You think you're doing the right thing. From what you know of the natural world, it should work. But then you do it and it completely backfires and something else happens."

"The real world of climate change is that we're up against it," Schrag responded. "Geoengineering is not something to do lightly. The reason we're thinking about it is because the real world has dealt us a shitty hand."

"We dealt it ourselves," Macfarlane said.

3

Right around the time the U.S. Navy launched Project Stormfury, the Army embarked on a project that was known—though only to a few, since it was top secret—as Iceworm. Project Iceworm was an exceptionally cold plan to win the Cold War. The Army proposed boring hundreds of miles of tunnels into Greenland's ice sheet. These would be outfitted with rail lines, and nuclear missiles would be shuttled along the tracks to keep the Soviets guessing. "Iceworm thus couples mobility with dispersion, concealment, and hardness," a classified report boasted.

Pursuant to this plan, in the summer of 1959 the Army Corps of Engineers was dispatched to build a base. Situated at seventy-seven degrees north latitude, about a hundred and fifty miles east of Baffin Bay, Camp Century was by far the biggest thing ever erected on—or within—the ice sheet. Using what were essentially giant snowblowers, the Corps dug a network of subsurface passages, which connected dorms, a mess hall, a chapel, a movie theater, and a barbershop. There was even a subglacial dispensary that sold perfume to send back home. (A favorite camp joke was there was a girl behind every tree.) Powering the enterprise was a portable nuclear reactor.

Camp Century was the one part of Project Iceworm the Army advertised. The base, it maintained, had been built to conduct polar research, and the Army produced a promotional film chronicling the herculean effort made by the Corps. Getting construction materials in from the coast required convoys of special tractors that labored across the ice at two miles an hour. "Camp Century is a symbol of man's unceasing struggle to conquer his environment," the narrator of the film intoned. Reporters were taken on tours through the tunnels, and two Boy Scouts—one American, one Danish—were invited north for a stay.

No sooner had construction been completed, though, than Camp Century's troubles began. Ice, like water, flows. The Corps knew this and had built the dynamic into its calculations. But the Corps hadn't adequately factored in the human factor—the way heat from the reactor would speed up the process. Almost at once, the tunnels started to contract. To keep the dorms, the movie theater, and the mess hall from being crushed, crews had to continually "trim" the ice with chainsaws. One visitor to the base compared the racket to the annual general meeting of all the devils of hell. By 1964, the chamber housing the reactor had deformed so much, the unit had to be removed. In 1967, the whole base was abandoned.

One way to gloss the Camp Century story is as another Anthropocene allegory. Man sets out to "conquer his environment." He congratulates himself for his resourcefulness and derring-do, only to find the walls closing in. Drive out nature with a snowblower, yet she will always hurry back.

But that's not the reason I'm telling it. Or at least not the main reason.

Camp Century may have been a Potemkin research station; still, actual research was conducted there. Even as the tunnels warped and buckled, a team of glaciologists set about drilling straight down through the ice sheet. The drilling team pulled up long, skinny cylinders of ice and kept going until they hit bedrock. The cylinders—more than a thousand in all—constituted the first complete Greenland ice core. What it revealed about the history of the climate was so puzzling and unlikely that scientists are still trying to make sense of it.

I first read about Camp Century when I was planning a trip of my own to Greenland. I had arranged to visit a Danish-led drilling operation called the North Greenland Ice Core Project, or North GRIP for short. The operation was situated on top of two miles of ice, in a spot even more remote than Camp Century. To get there, I hitched a ride on a ski-equipped C-130 Hercules, which those in the know call a Herc. The flight was carrying several thousand feet of drilling cable, a team of European glaciologists, and Denmark's minister of research. (Greenland is a Danish territory, a fact the U.S. Army cheerfully ignored in planning for Iceworm.) Like the rest of us, the minister had to sit in the Herc's hold, wearing military-issue earplugs.

One of North GRIP's directors, J. P. Steffensen, greeted us when we disembarked. We were dressed in huge insulated boots and heavy snow gear. Steffensen had on a pair of old sneakers, a

One of the entrances to Camp Century

filthy parka that was flapping open, and no gloves. Tiny icicles hung from his beard. First he delivered a short lecture on the dangers of dehydration. "It sounds like a complete contradiction in terms," he told us. "You're standing on three thousand meters of water. But it's extremely dry. So make sure you have to go and pee." Then he briefed us on camp protocol. There were two frost-proof toilets from Sweden, but men were kindly requested to relieve themselves out on the ice, at a spot designated by a little red flag.

North GRIP was a decidedly modest affair. It consisted of a

Camp Century's tunnels had to be maintained with chainsaws.

half dozen cherry-red tents arrayed around a geodesic dome that had been purchased, mail order, from Minnesota. In front of the dome, someone had planted the standard jokey symbol of isolation—a milepost showing the nearest town, Kangerlussuaq, to be five hundred miles away. Nearby stood the standard jokey symbol of the cold—a plywood palm tree. The view in all directions was exactly the same: an utterly flat expanse of white that could be described as bleak or, alternatively, as sublime.

Beneath the camp, an eighty-foot-long tunnel led down to the drilling room. This chamber had been hollowed out of the ice, like the passageways at Camp Century, and inside it, the temperature, even in June, never rose above freezing. Again as at Camp Century, the chamber was shrinking. Pine beams had

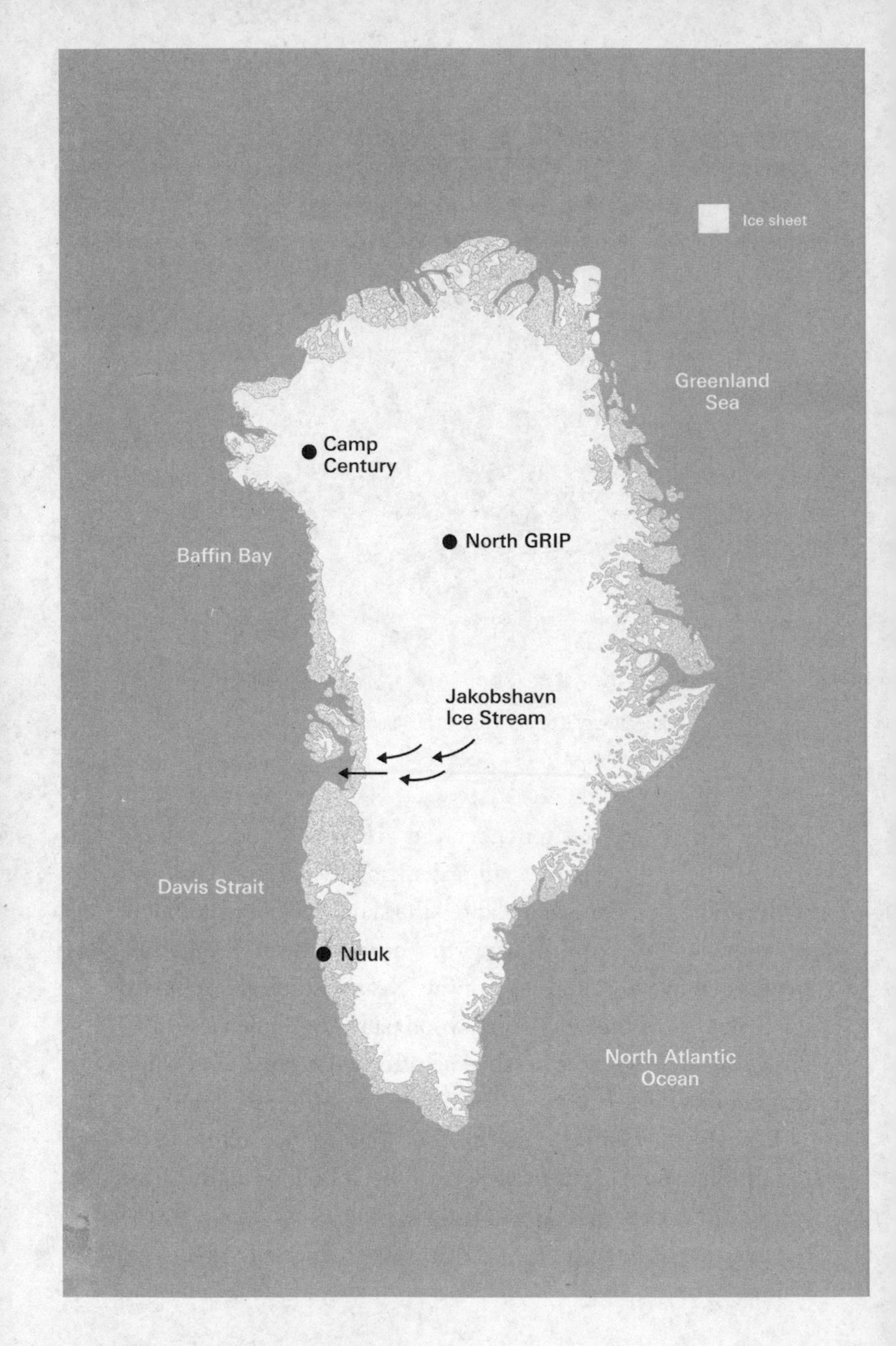
Ice sheet
Greenland
Sea
Camp
Century
North GRIP
Baffin Bay
Jakobshavn
Ice Stream
Davis Strait
Nuuk
North Atlantic
Ocean

been installed to reinforce the ceiling, but they'd already shattered under the weight of the snow. Drilling began every morning at 8 A.M. The first task of the day was to lower the drill, a twelve-foot-long tube with fierce metal teeth on one end, down to the bottom of the borehole. Once in position, the toothy tube was set spinning, so that an ice cylinder gradually formed within it. The cylinder was then pulled up by means of a steel cable. The first time I watched the process, a glaciologist from Iceland and another from Germany were manning the controls. At the depth they had reached—nine thousand six hundred and eighty feet—it took an hour just for the drill to descend. During that time, there wasn't much for the pair to do except watch their computers, which sat on little heating pads, and listen to ABBA. "The word 'stuck' is not in our vocabulary," the Icelander told me, with a nervous laugh.

Like all glaciers, the Greenland ice sheet is made up entirely of accumulated snow. The most recent layers are thick and airy, while the older layers are thin and dense, which means that to drill down through the ice is to descend backward in time, at first gradually and then much more rapidly. About a hundred and forty feet down, there's snow dating from the American Civil War; some twenty-five hundred feet down, snow from the time of Plato; and at a depth of five thousand three hundred and fifty feet, snow from when prehistoric painters were decorating the caves at Lascaux. As the snow is compressed, its crystal structure changes to ice. But in most other respects, it remains unchanged, a relic of the moment it formed. In the Greenland ice, there's volcanic ash from Tambora, lead pollution from Roman smelters, and dust blown in from Mongolia on ice age winds. Every layer contains tiny bubbles of trapped air, each a sample of a past atmosphere. To someone who knows how to read them, the layers are an archive of the sky.

Eventually, the drill team pulled up a short section of core—about two feet long and four inches in diameter. Someone went to fetch the minister, who arrived in the chamber wearing a red snowsuit. The section looked a lot like a two-foot-long cylinder of ordinary ice. But, one of the drillers explained, it was made up of snow that had fallen over a hundred and five thousand years ago, at the beginning of the last ice age. The minister exclaimed something in Danish and seemed suitably impressed.

The first person to realize just how much information could be gleaned from an ice core was a geophysicist named Willi Dansgaard. Dansgaard, who was also Danish, was an expert on the chemistry of precipitation. Presented with a sample of rainwater, he could, based on its isotopic composition, determine the temperature at which it had formed. This method, he realized, could also be applied to snow. When Dansgaard heard about the Camp Century core, in 1966, he applied for permission to analyze it. He was more than a little surprised when it was granted. The Americans, he later wrote, didn't seem to realize what a "gold mine" of data they had in their refrigerated vault.

In its broad outlines, Dansgaard's reading of the Camp Century core confirmed what was already known about climate history. The most recent ice age, known in the United States as the Wisconsin, began roughly a hundred and ten thousand years ago. During the Wisconsin, ice sheets spread over the northern hemisphere until they covered Scandinavia, Canada, New England, and much of the upper Midwest. Throughout this period, Greenland was frigid. When the Wisconsin ended, roughly ten thousand years ago, Greenland (and the rest of the world) warmed.

The details were a different matter. Dansgaard's analysis of the core suggested that in the midst of the last ice age, the climate of Greenland was so variable it could hardly be called *a* climate. Average temperatures on the ice sheet had, it appeared, shot up by as much as 8°C—more than 14°F—in fifty years. Then they had dropped again, almost as abruptly. This had happened not just once but many times. A temperature swing of 14°F? It was as if New York City had suddenly become Houston, or Houston had become Riyadh, and then flipped back again. Everyone, including Dansgaard, was perplexed. Could these violent swings in the data correspond to real events? Or did they represent some kind of glitch?

Over the next four decades, five more cores were extracted from different parts of the ice sheet. Each time, the wild swings showed up. Meanwhile, other climate records, including pollen deposits from a lake in Italy, ocean sediments from the Arabian Sea, and stalagmites from a cave in China, revealed the same pattern. The temperature swings became known, after Dansgaard and a Swiss colleague, Hans Oeschger, as Dansgaard–Oeschger events. There are twenty-five such D–O events recorded in the Greenland ice. Richard Alley, a glaciologist at Penn State, has compared the effect to watching "a three-year-old who has just discovered a light switch, flicking it back and forth."

The last great swing took place as the ice age was ending, and it was a doozy. Temperatures in Greenland shot up by 15°F in a decade, or perhaps even faster. Then things settled into a new and very different regime. For the next ten thousand years, temperatures in Greenland (and the rest of the world) remained more or less constant, decade after decade, century after century.

All of civilization falls within this period of relative tranquility, and so this sort of calm is what we take to be the norm. It's an understandable mistake, but still a mistake. Over the last hun-

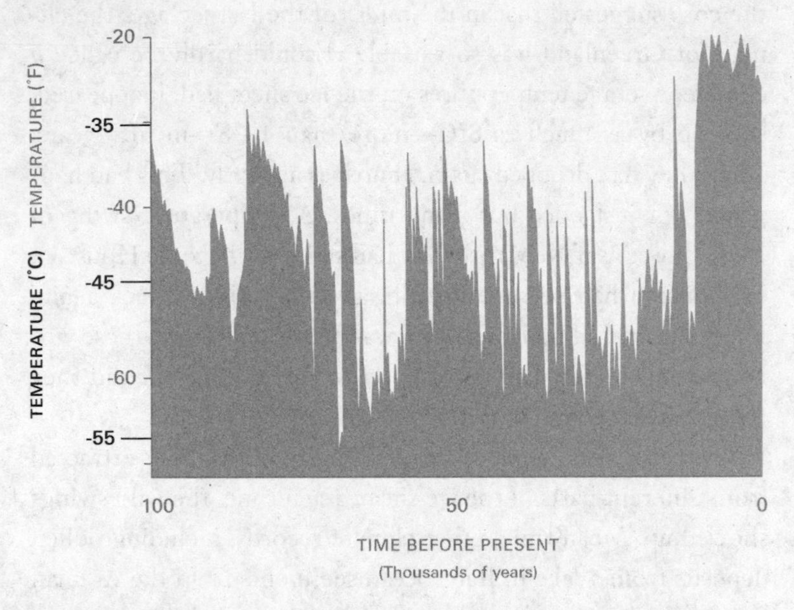

During the last ice age, temperatures over central Greenland swung wildly.

dred and ten thousand years, the only period as stable as our own *is* our own.

One night at North GRIP, I interviewed Steffensen in the geodesic dome. It was midnight but polar day, so outside the sun was shining. The glaciologists were drinking beer, playing board games, and listening to the soundtrack from *Buena Vista Social Club*.

I brought up the issue of climate change. Perhaps, I suggested hopefully, it would ward off another ice age and more D–O events. At least we could dodge that particular disaster!

Steffensen was unimpressed by my suggestion. He pointed out that if you believed the climate to be inherently unstable, the last thing you'd want to do is mess around with it. He recited an old Danish saying, whose pertinence I didn't entirely understand but which nonetheless stuck with me. He translated it as, "Pissing in your pants will only keep you warm for so long."

We got to talking about climate history and human history. In Steffensen's view, these amounted to more or less the same thing. "If you look at the output of ice cores, it has really changed the picture of the world, our view of past climates and of human evolution," he told me. "Why did human beings not make civilization fifty thousand years ago?

"You know that they had just as big brains as we have today," he went on. "When you put it in a climatic framework, you can say, well, it was the ice age. And also this ice age was so climatically unstable that each time you had the beginnings of a culture, they had to move. Then comes the present interglacial—ten thousand years of very stable climate. The perfect conditions for agriculture. If you look at it, it's amazing. Civilizations in Persia, in China, and in India start at the same time, maybe six thousand years ago. They all developed writing and they all developed religion and they all built cities, all at the same time, because the climate was stable. I think that if the climate would have been stable fifty thousand years ago, it would have started then. But they had no chance."

I was contemplating another trip to Greenland, where Steffensen and his colleagues were drilling a new ice core, when COVID-19 hit. Suddenly everyone's plans were upended, including my own. As borders closed and flights were canceled, travel to the ice sheet—or, for that matter, pretty much anywhere—became impractical. Here I was, trying to finish a book about the world spinning out of control, only to find the world spinning so far out of control that I couldn't finish the book.

Scientists are still trying to puzzle out what caused the wild temperature swings first glimpsed in the Camp Century core. One hypothesis is that they are related to a loss of sea ice in the Arctic, which is worrisome, given that global warming is causing a loss of sea ice in the Arctic. But even putting aside the possibility of a human-induced D–O event, the calm of the last ten thousand years is clearly coming to an end. Without intending to, or even realizing it, humanity has used the stability it lucked into to create Greenland-scale instability.

Since 1990, temperatures on the ice sheet have risen by almost 3°C (more than 5°F). During the same period, ice loss from Greenland has increased sevenfold, from thirty billion tons a year to an average of more than two hundred fifty billion tons a year. Melt is occurring over more and more area and at higher and higher elevations: during an exceptionally warm couple of days in the summer of 2019, melting was detected on more than ninety-five percent of the ice sheet's surface. That summer—a record-breaker—Greenland shed almost six hundred billion tons of ice, producing enough water to fill a pool the size of California to a depth of four feet.

"The current Arctic is experiencing rates of warming comparable to abrupt changes, or D–O events, recorded in Greenland ice cores," a team of Danish and Norwegian scientists recently reported. Since the melt process is self-reinforcing—water is dark and absorbs sunlight, while ice is light-colored and reflects it—there's widespread concern that Greenland may be approaching the point beyond which the disintegration of the entire ice sheet becomes inevitable. This could take centuries—even millennia—to play out, but, all told, there's enough ice on Greenland to raise global sea levels by twenty feet.

As with temperatures, sea levels have in the past varied dramatically. At the end of the Wisconsin, as the great ice sheets were breaking up, there were periods when they rose at the astonishing rate of a foot a decade. (It's been proposed that one of these "meltwater pulses" inspired the account of the deluge in Genesis.) Obviously, our ancestors dealt with this tumult, or we wouldn't be here. But, in contrast to us, they traveled light. How—and where—would you relocate a city like Boston or Mumbai or Shenzhen? Private ownership, national boundaries, subway lines, transmission cables, sewage pipes—all these are relatively recent developments in human society, and they all militate against picking up and moving. In this sense, just about every coastal city is, like New Orleans, committed to stasis and to the costly and increasingly elaborate interventions that maintaining stasis will require. To combat rising sea levels and the more deadly storm surges that they bring, the Army Corps of Engineers has proposed building a series of artificial islands in New York Harbor. These would be connected by six miles of huge retractable gates. An early cost estimate for the project ran to more than $100 billion. Alternatively, it's been proposed that sea-level rise could be slowed by propping up Antarctic ice shelves or by blocking the mouth of one of Greenland's largest outlet glaciers, the Jakobshavn ice stream.

"We understand the hesitancy to interfere with glaciers," the authors of this proposal—scientists from the United States and Finland—observed in *Nature*. "As glaciologists, we know the pristine beauty of these places." But "if the world does nothing, ice sheets will keep shrinking and the losses will accelerate. Even if greenhouse-gas emissions are slashed, which looks unlikely, it would take decades for the climate to stabilize."

First you speed up an ice stream; then you try to slow it down by erecting a three-hundred-foot-tall, three-mile-long concrete-topped embankment.

• • •

This has been a book about people trying to solve problems created by people trying to solve problems. In the course of reporting it, I spoke to engineers and genetic engineers, biologists and microbiologists, atmospheric scientists and atmospheric entrepreneurs. Without exception, they were enthusiastic about their work. But, as a rule, this enthusiasm was tempered by doubt. The electric fish barriers, the concrete crevasse, the fake cavern, the synthetic clouds—these were presented to me less in a spirit of techno-optimism than what might be called techno-fatalism. They weren't improvements on the originals; they were the best that anyone could come up with, given the circumstances. As one replicant in *Blade Runner* says to Harrison Ford, who may or may not be playing a replicant: "You think I'd be working in a place like this if I could afford a real snake?"

It's in this context that interventions like assisted evolution and gene drives and digging millions of trenches to bury billions of trees have to be assessed. Geoengineering may be "entirely crazy and quite disconcerting," but if it could slow the melting of the Greenland ice sheet, or take some of "the pain and suffering away," or help prevent no-longer-fully-natural ecosystems from collapsing, doesn't it have to be considered?

Andy Parker is the project director for the Solar Radiation Management Governance Initiative, which works to expand the "global conversation" around geoengineering. His preferred drug analogy for the technology is chemotherapy. No one in his right mind would undergo chemotherapy were better options available. "We live in a world," he has said, "where deliberately dimming the fucking sun might be less risky than not doing it."

But to imagine that "dimming the fucking sun" could be less dangerous than not dimming it, you have to imagine not only that the technology will work according to plan but also that it will be deployed according to plan. And that's a lot of imagining. As Keutsch, Keith, and Schrag all pointed out to me, scientists can only make recommendations; implementation is a political decision. You might hope that such a decision would be made equitably with respect to those alive today and to future generations, both human and nonhuman. But let's just say the record here isn't strong. (See, for example, climate change.)

Suppose that the world—or just a small group of assertive nations—launched a fleet of SAILs. And suppose that even as the SAILs are flying and lofting more and more tons of particles, global emissions continue to rise. The result would not be a return to the climate of pre-industrial days or to that of the Pliocene or even that of the Eocene, when crocodiles basked on Arctic shores. It would be an unprecedented climate for an unprecedented world, where silver carp glisten under a white sky.

Afterword

Endings matter. This isn't just a truism; it's a scientific fact demonstrated in dozens of experiments, perhaps the most compelling of which involved colonoscopies. About thirty years ago, researchers divided colonoscopy patients at a Toronto hospital into two groups. Those in the first underwent the procedure with only a mild sedative. Those in the second got the same mild sedative as well as something extra. After the medically necessary part of the exam was completed, the physicians left the tip of the colonoscope inside the patients' rectums for an additional three minutes. This obviously lengthened the exam's duration, but, as the researchers noted, it also "resulted in final moments that were less painful." Later, all of the patients were asked to reflect on their experience. Those in the second group, who'd actually suffered for more time, rated the procedure as less unpleasant than those in the first. "Last impressions may be lasting impressions," the researchers concluded.

Recent books about the environment, which tend to be if not exactly painful then at least gloomy, usually end with what I will call the additional three pages. After describing just how bad things really are, for insects or orangutans or ice caps or the planet in general, the author explains why there's reason for optimism. Often the final chapter includes steps for the concerned reader to take—plant native flowers, ride a bike, take to the streets, or, if all else fails, decamp to Mars. The last impression left is, if not exactly upbeat, then at least less unpleasant.

I'd like to end that way, too. But I can't.

The preceding pages were written, perhaps not coincidentally, in a period that roughly coincided with the administration of Donald J. Trump. (The hardcover publication date was the first day of Trump's second impeachment trial.) Since then, a great deal has happened, including coups in Myanmar and Mali, the fall of Afghanistan to the Taliban, a record-shattering heat wave in the Pacific Northwest, and the first-ever water shortage declaration on the Colorado River. All the while, the world has remained preoccupied with COVID.

COVID both interrupted this book and belongs within it. Like the melting of the Greenland ice sheet, the bleaching of the Great Barrier Reef, and the sinking of southern Louisiana, it's a manmade natural disaster. It is a microbial version of a CHANS, a coupled human and natural system. It's nothing that anyone intended; still, it is something we brought about—a product of an experiment we are running on the world and on ourselves.

Let's assume COVID was the result of a "spillover" event—an evolutionary accident that enabled the virus, SARS-CoV-2, to jump from a species of bat, or, more likely, an intermediary species that had caught it from a bat, into humans. When the first cases were reported in Wuhan, in the final days of 2019, the virus probably could have been contained, had the proper protocols been followed. But politics intervened. Chinese officials temporized, censoring reports about the virus and accusing physicians in Wuhan of "spreading rumors."

Wuhan, a metropolis of eleven million, is a center of global commerce, linked to the rest of the world by bullet train and jet. Within a month of the first confirmed cases, COVID had turned up in two dozen countries, including Italy, Germany, Russia, Australia, Malaysia, and the United States. Soon, it was everywhere—Greenland, the Falklands, the Kuril Islands off the Kamchatka Peninsula.

COVID's spread took the world by surprise, but, really, it shouldn't have. Epidemiologists have been warning for decades of just such an eventuality. As David Quammen noted in his book *Spillover*, published in 2012, contemporary culture is practically a recipe for a pandemic:

> We live at high densities in many cities. We have penetrated, and we continue to penetrate, the last great forests and other wild ecosystems of the planet . . . We settle in those places, creating villages, work camps, towns, extractive industries, new cities. We bring in our domesticated animals, replacing wild herbivores with livestock. We multiply our livestock as we've multiplied ourselves, operating huge factory-scale operations involving thousands of cattle, pigs, chickens, ducks, sheep, and goats . . . all confined en masse within pens and corrals, under conditions that allow those domestics and semidomestics to acquire infectious pathogens from external

> sources (such as bats roosting over the pig pens) . . . We travel, moving between cities and continents even more quickly than our transported livestock.

As soon as it became clear that COVID couldn't be contained, work began on a technofix, in this case a bio-technofix. More than ninety teams rushed to develop a vaccine. In record time, several succeeded. Two of the vaccines created, by Moderna and Pfizer, involved a cutting-edge technique featuring messenger RNA. Briefly it seemed that the technofix might offer a way out; then the much less rosy reality set in. As I write this, hospitals in the United States are once again filled with COVID patients, some suffering from so-called "breakthrough" infections. The U.S. government is preparing to administer booster shots, even as billions of people around the globe remain unvaccinated, and scientists are warning that the virus is now endemic, meaning it will continue to circulate pretty much no matter what. At this point, trying to eradicate COVID is "like trying to plan the construction of a stepping-stone pathway to the Moon," is how one epidemiologist put it to the journal *Nature.*

All of this is not to say there's nothing to be done. Planting native flowers, riding bikes, installing solar panels—these are all steps that, if taken by enough people, would make a real difference. Political action is essential, and protest is one way to effect it. But in this book, I've tried to be honest about the scale of the challenge and about the difficulty of arresting processes of global change once they've been set in motion. It would be disingenuous for me to use these final pages to suggest otherwise.

One last point: neither optimism nor pessimism alters the fact that we live in an extraordinary moment. The choices that we'll make—that we are making right now, without necessarily being aware of them—will determine the future of life for our children and their children and all the other species on earth for generations to come. This is the situation we're in, and there's no avoiding it, because, in the end, this marvelous, fragile planet is all we've got.

October 2021

Acknowledgments

This book could not have been written without a lot of help. I am deeply grateful to the many people who shared with me their expertise, their experiences, and their time.

For help understanding how Asian carp got to the United States and where they're going, I'd like to thank Margaret Frisbie, Mike Alber, and the Friends of the Chicago River, who took me on a wonderful adventure on *City Living*. I also want to thank Chuck Shea, Kevin Irons, Philippe Parola, Clint Carter, Duane Chapman, Robin Calfee, Anita Kelly, Drew Mitchell, and Mike Freeze. Thanks, too, to Tracy Seidemann and the Illinois DNR biologists and contract fishermen who put up with me and my endless questions.

Owen Bordelon kindly (and expertly) flew me over Plaquemines Parish, and David Muth and Jacques Hebert helped make that happen. Clint Willson, Rudy Simoneaux, Brad Barth, Alex Kolker, Boyo Billiot, Chantel Comardelle, Jeff Hebert, Joe Harvey, and Chuck Perrodin were all great guides to the complexities of life along the Mississippi.

The people working to keep the desert fishes of the United States alive deserve a special kind of gratitude. Thanks to Kevin Wilson, Jenny Gumm, Olin Feuerbacher, Ambre Chaudoin, Jeff Goldstein, and Brandon Senger, who took me pupfish-counting at Devils Hole. Thanks, too, to Kevin Guadalupe, who showed me Nevada's poolfish and without whom there might not have been any to show, and to Susan Sorrells, who has worked so hard to keep the Shoshone pupfish alive. I am grateful, too, to Kevin Brown, who shared with me his report on the history of Devils Hole.

Ruth Gates passed away when I was midway through this book. I feel very fortunate to have been able to spend time with her on Moku o Lo'e and for her help when I was just beginning to conceive of this project. I am also extremely grateful to Madeleine van Oppen and to all of the other dedicated marine scientists I met when I was in Australia, including Kate Quigley, David Wachenfeld, Annie Lamb, Patrick Buerger, and Wing Chan. Thanks, too, to Paul Hardisty and Marie Roman.

Mark Tizard and Caitlin Cooper were incredibly generous to me when I visited them in Geelong. Paul Thomas was equally so when I went to visit him in Adelaide. Genetic engineering is an immensely complicated topic, and I thank all three of them for so patiently explaining their work to me. Lin Schwarzkopf very kindly took me toad hunting. Thanks to Royden Saah at GBIRd, and many thanks to Luana Maroja, at Williams College, who generously helped me with the finer points of gene drive.

I was very lucky to visit the Hellisheiði Power Station with Edda Aradóttir despite the restrictions imposed by COVID. Thanks to her and also to Ólöf Baldursdóttir for making that happen. Klaus Lackner was a wonderful host when I met with him at ASU. Jan Wurzbacher, Louise Charles, and Paul Ruser were generous with their time when I visited Zurich. Gratitude to Oliver Geden, Zeke Hausfather, and Magnús Bernhardsson.

I went to talk to Frank Keutsch, David Keith, and Dan Schrag at Harvard just a few days before the entire campus shut down due to COVID. I want to thank them all for taking the time to walk me through the many complexities—both technical and ethical—of solar geoengineering. Thanks to Allison Macfarlane, who, in a very real sense, walked onto these pages, and also to Lizzie Burns, Zhen Dai, Sir David King, Andy Parker, Gernot Wagner, Janos Pasztor, and Cynthia Scharf.

In a roundabout sort of way, this book owes its origins to the visit I paid to North GRIP when it still existed. Thanks to J. P. Steffensen, Dorthe Dahl-Jensen, Richard Alley, and the many intrepid glaciologists who are working to understand the past and the future of the Greenland ice sheet. Thanks, too, to Ned Kleiner, my favorite climate scientist, who read and commented on key chapters, and to Aaron and Matthew Kleiner, who offered crucial last-minute advice.

I am grateful to the Alfred P. Sloan Foundation for its generous assistance. A grant from the foundation supported research and travel for this book and allowed me to report from places I otherwise might not have been able to go. In 2019, I spent a month working on this project at the Rockefeller Foundation's Bellagio Center. The setting was amazing and the company inspiring. Parts of this book were also written while I was a fellow at the Williams College Center for Environmental Studies. A shout-out to the students and faculty at CES. A special thanks to Walton Ford, whose great auk provided inspiration in dark times.

Many people worked under a tight deadline to turn the manuscript I submitted into a book. Heartfelt thanks to Caroline Wray, Simon Sullivan, Evan Camfield, Kathy Lord, Janice Ackerman, Alicia Cheng, Sarah Gephart, Ian Keliher, and the team at MGMT Design. I am indebted to Julie Tate, who fact-checked several of these chapters, and to the fact-checking team at *The New Yorker*. Any errors that remain are entirely my own.

Sections of this book first appeared in *The New Yorker*. I am profoundly grateful to David Remnick, Dorothy Wickenden, John Bennet, and Henry Finder for their counsel and support over lo these many years.

Gillian Blake never lost faith in this project despite the complexities that arose along the way. I can't thank her enough for her encouragement, her editorial advice, and her good judgment. Kathy Robbins was, as always, a great friend. An author could not ask for a more discerning reader or a more stalwart advocate.

Finally, I want to thank my husband, John Kleiner. To borrow from Darwin, this book came half out of his brain, and I'm not sure how to acknowledge this sufficiently "without saying so in so many words." Not a single page of this would have been written without his insight, his enthusiasm, and his willingness to read yet another draft.

Notes

Down the River

1

3 **"the grimmest and most dead-earnest of reading matter":** Mark Twain, *Life on the Mississippi*, reprint ed. (New York: Penguin Putnam, 2001), 54.

3 **"Going up that river":** Joseph Conrad, *Heart of Darkness and The Secret Sharer*, reprint ed. (New York: Signet Classics, 1950), 102.

5 **WATER IN CHICAGO RIVER:** *The New York Times* (Jan. 14, 1900), 14.

5 **became known as the Chicago School of Earth Moving:** Libby Hill, *The Chicago River: A Natural and Unnatural History* (Chicago: Lake Claremont Press, 2000), 127.

5 **an island more than fifty feet high and a mile square:** Cited in Hill, *The Chicago River*, 133.

7 **transformed more than half the ice-free land on earth:** Roger LeB. Hooke and José F. Martín-Duque, "Land Transformation by Humans: A Review," *GSA Today*, 22 (2012), 4–10.

7 **felt all the way in Des Moines:** Katy Bergen, "Oklahoma Earthquake Felt in Kansas City, and as Far as Des Moines and Dallas," *The Kansas City Star* (Sept. 3, 2016), kansascity.com/news/local/article99785512.html.

7 **"humans and livestock outweigh all vertebrates combined":** Yinon M. Bar-On, Rob Phillips, and Ron Milo, "The Biomass Distribution on Earth," *Proceedings of the National Academy of Sciences*, 115 (2018), 6506–6511.

8 **to disguise the project's true purpose:** "Historical Vignette 113—Hide the Development of the Atomic Bomb," U.S. Army Corps of Engineers Headquarters, usace.army.mil/About/History/Historical-Vignettes/Military-Construction-Combat/113-Atomic-Bomb/.

11 **The Corps considered more than a dozen:** P. Moy, C. B. Shea, J. M. Dettmers, and I. Polls, "Chicago Sanitary and Ship Canal Aquatic Nuisance Species Dispersal Barriers," report available for download at: glpf.org/funded-projects/aquatic-nuisance-species-dispersal-barrier-for-the-chicago-sanitary-and-ship-canal/.

12 **"ruin our way of life":** Quoted in Thomas Just, "The Political and Economic Implications of the Asian Carp Invasion," *Pepperdine Policy Review*, 4 (2011), digitalcommons.pepperdine.edu/ppr/vol4/iss1/3.

13 **"the first documented example of integrated polyculture":** Patrick M. Kočovský, Duane C. Chapman, and Song Qian, "'Asian Carp' Is Societally and Scientifically Problematic. Let's Replace It," *Fisheries*, 43 (2018), 311–316.

13 **almost fifty billion pounds in 2015 alone:** Figures from the *China Fisheries Yearbook 2016*, cited in Louis Harkell, "China Claims 69m Tons of Fish Produced in 2016," *Undercurrent News* (Jan. 19, 2017), undercurrentnews.com/2017/01/19/ministry-of-agriculture-china-produced-69m-tons-of-fish-in-2016/.

14 **whose working title was *The Control of Nature*:** William Souder, *On a Farther Shore: The Life and Legacy of Rachel Carson* (New York: Crown, 2012), 280.

14 **"The 'control of nature' is a phrase conceived in arrogance":** Rachel Carson, *Silent Spring*, 40th anniversary ed. (New York: Mariner, 2002), 297.

15 **the first documented shipment of Asian carp:** Andrew Mitchell and Anita M. Kelly, "The Public Sector Role in the Establishment of Grass Carp in the United States," *Fisheries*, 31 (2006), 113–121.

15 **the Arkansas Game and Fish Commission found a use:** Anita M. Kelly, Carole R. Engle, Michael L. Armstrong, Mike Freeze, and Andrew J. Mitchell, "History of Introductions and Governmental Involvement in Promoting the Use of Grass, Silver, and Bighead Carps," in *Invasive Asian Carps in North America*, Duane C. Chapman and Michael H. Hoff, eds. (Bethesda, Md.: American Fisheries Society, 2011), 163 171.

17 **"Who hears the fishes when they cry?":** Henry David Thoreau, *A Week on the Concord and Merrimack Rivers*, reprint ed. (New York: Penguin, 1998), 31.

17 **Bigheads can, on occasion, weigh as much as a hundred pounds:** Duane C. Chapman, "Facts About Invasive Bighead and Silver Carps," publication of the United States Geological Survey, available at: pubs.usgs.gov/fs/2010/3033/pdf/FS2010-3033.pdf.

18 **"Bighead and silver carp don't just invade ecosystems":** Dan Egan, *The Death and Life of the Great Lakes* (New York: Norton, 2017), 156.

18 **on some waterways the proportion is even higher:** Dan Chapman, *A War in the Water*, U.S. Fish and Wildlife Service, southeast region (March 19, 2018), fws.gov/southeast/articles/a-war-in-the-water/.

20 **The result was fifty-four thousand pounds of dead fish:** Egan, *The Death and Life of the Great Lakes*, 177.

20 **"no greater threat to the ecosystem of the Great Lakes":** Cited in Tom Henry, "Congressmen Urge Aggressive Action to Block Asian Carp," *The Blade* (Dec. 21, 2009), toledoblade.com/local/2009/12/21/Congressmen-urge-aggressive-action-to-block-Asian-carp/stories/200912210014.

20 **Michigan filed a lawsuit:** "Lawsuit Against the U.S. Army Corps of Engineers and the Chicago Water District," Department of the Michigan Attorney General, michigan.gov/ag/0,4534,7-359-82915_82919_82129_82135-447414--,00.html.

21 **According to the Corps' assessment:** The Great Lakes and Mississippi River Interbasin Study, or GLMRIS report, is available at: glmris.anl.gov/glmris-report/.

21 **On the Great Lakes side:** A list of the (at last count) 187 invasive species established in the Great Lakes is provided by NOAA at: glerl.noaa.gov/glansis/GLANSISposter.pdf.

22 **A woman I read about:** Phil Luciano, "Asian Carp More Than a Slap in the Face," *Peoria Journal Star* (Oct. 21, 2003), pjstar.com/article/20031021/NEWS/310219999.

26 **the *China Daily* ran an article:** Doug Fangyu, "Asian Carp: Americans' Poison, Chinese People's Delicacy," *China Daily USA* (Oct. 13, 2014), http://usa.chinadaily.com.cn/epaper/2014-10/13/content_18730596.htm.

2

31 **officially retired thirty-one Plaquemines place names:** Amy Wold, "Washed Away: Locations in Plaquemines Parish Disappear from Latest NOAA Charts," *The Advocate* (Apr. 29, 2013), theadvocate.com/baton_rouge/news/article_f60d4d55-e26b-52c0-b9bb-bed2ae0b348c.html.

32 **"We harnessed it, straightened it, regularized it, shackled it":** Cited in John McPhee, *The Control of Nature* (New York: Noonday, 1990), 26.

33 **some four hundred million tons' worth annually:** Liviu Giosan and Angelina M. Freeman, "How Deltas Work: A Brief Look at the Mississippi River Delta in a Global Context," in *Perspectives on the Restoration of the Mississippi Delta*, John W. Day, G. Paul Kemp, Angelina M. Freeman, and David P. Muth, eds. (Dordrecht, Netherlands: Springer, 2014), 30.

35 **had been assured by a Bayogoula guide:** Christopher Morris, *The Big Muddy: An Environmental History of the Mississippi and Its Peoples from Hernando de Soto to Hurricane Katrina* (Oxford: Oxford University Press, 2012), 42.

35 **wading "mid-leg deep" to get to their cabins:** Cited in Morris, *The Big Muddy*, 45.

35 **"I do not see how settlers can be placed on this river":** Cited in Morris, *The Big Muddy*, 45.

35 **"The site is drowned under half a foot of water":** Cited in Lawrence N. Powell, *The Accidental City: Improvising New Orleans* (Cambridge, Mass.: Harvard University Press, 2012), 49.

36 **slave-built levees stretched along both banks:** Morris, *The Big Muddy*, 61.

36 **extended for more than a hundred and fifty miles:** John M. Barry, *Rising Tide: The Great Mississippi Flood of 1927 and How It Changed America* (New York: Touchstone, 1997), 40.

44 **In 1735, a crevasse-induced flood:** Donald W. Davis, "Historical Perspective on Crevasses, Levees, and the Mississippi River," in *Transforming New Orleans and Its Environs*, Craig E. Colten, ed. (Pittsburgh: University of Pittsburgh, 2000), 87.

44 **observed "one sheet of water":** Cited in Richard Campanella, "Long before Hurricane Katrina, There Was Sauve's Crevasse, One of the Worst Floods in New Orleans History," *nola.com* (June 11, 2014), nola.com/entertainment_life/home_garden/article_ea927b6b-d1ab-5462-9756-ccb1acdf092e.html.

44 **In 1858, forty-five crevasses opened up:** For a full account of crevasses, 1773–1927, see Davis, "Historical Perspectives on Crevasses, Levees, and the Mississippi River," 95.

44 **two hundred and twenty-six crevasses were reported:** Davis, "Historical Perspectives on Crevasses, Levees, and the Mississippi River," 100.

44 **caused an estimated $500 million worth of damage:** Estimates of the damages caused by the Great Flood of 1927 vary widely; some are as high as a billion dollars, or almost $15 billion in today's money.

45 **the most important piece of water-related legislation:** Cited in Christine A. Klein and Sandra B. Zellmer, *Mississippi River Tragedies: A Century of Unnatural Disaster* (New York: New York University, 2014), 76.

45 **within four years, it had added:** D. O. Elliott, *The Improvement of the Lower Mississippi River for Flood Control and Navigation: Vol. 2* (St. Louis: Mississippi River Commission, 1932), 172.

45 **On average, the levees were raised by three feet:** Elliott, *The Improvement of the Lower Mississippi River: Vol. 2*, 326.

45 **A poem commemorating the Corps' efforts:** The excerpt comes from Michael C. Robinson, *The Mississippi River Commission: An American Epic* (Vicksburg, Miss.: Mississippi River Commission, 1989).

46 **"The Mississippi River was controlled; land was lost":** Davis, "Historical Perspectives on Crevasses, Levees, and the Mississippi River," 85.

47 **The authority had taken them anyway:** John Snell, "State Takes Soil Samples at Site of Largest Coastal Restoration Project, Despite Plaquemines Parish Opposition," *Fox8live* (last updated Aug. 23, 2018), fox8live.com/story/38615453/state-takes-soil-samples-at-site-of-largest-coastal-restoration-project-despite-plaquemines-parish-opposition/.

49 **dropping by almost half a foot a decade:** Cathleen E. Jones et al., "Anthropogenic and Geologic Influences on Subsidence in the Vicinity of New Orleans, Louisiana," *Journal of Geophysical Research: Solid Earth*, 121 (2016), 3867–3887.

49 **"New Orleans' drainage problem is a terrible one":** Thomas

Ewing Dabney, "New Orleans Builds Own Underground River," *New Orleans Item* (May 2, 1920), 1.

51 **THE CASE AGAINST REBUILDING THE SUNKEN CITY OF NEW ORLEANS:** Jack Shafer, "Don't Refloat: The Case against Rebuilding the Sunken City of New Orleans," *Slate* (Sept. 7, 2005), slate.com/news-and-politics/2005/09/the-case-against-rebuilding the-sunken-city-of-new-orleans.html.

51 **"It is time to face up to some geological realities":** Klaus Jacob, "Time for a Tough Question: Why Rebuild?" *The Washington Post* (Sept. 6, 2005).

51 **An advisory group appointed by New Orleans's mayor:** Reports of the Bring New Orleans Back Commission, appointed by Mayor Ray Nagin, are archived at: columbia.edu/itc/journalism/cases/katrina/city_of_new_orleans_bnobc.html.

52 **twelve thousand cubic feet of water per second:** Mark Schleifstein, "Price of Now-Completed Pump Stations at New Orleans Outfall Canals Rises by $33.2 Million," *New Orleans Times-Picayune* (last updated July 12, 2019), nola.com/news/environment/article_7734dae6-c1c9-559b-8b94-7a9cef8bb6d8.html.

52 **twenty miles closer to the Gulf:** Klein and Zellmer, *Mississippi River Tragedies*, 144.

52 **for every three miles a storm has to travel:** How much wetlands buffer storm surges is a much-debated topic. This estimate is cited in Klein and Zellmer, *Mississippi River Tragedies*, 141.

54 **Jean Marie's children, in turn, married descendants:** The history of the Isle de Jean Charles Band of the Biloxi-Chitimacha-Choctaw Tribe, as well as the latest on the resettlement plan, can be found at isledejeancharles.com.

55 **the project's billion-dollar price tag:** The price of the Morganza to the Gulf project keeps changing. These figures come from the late 1990s, when the Corps decided not to include Isle de Jean Charles inside the levees.

58 **"The Corps of Engineers can make the Mississippi River go":** McPhee, *The Control of Nature*, 50.

58 **"The word will now come to mind":** McPhee, *The Control of Nature*, 69.

Into the Wild

1

63 **not far from Mount Stirling:** In Manly's day, the mountain had not been officially named; his location is reckoned in Richard E. Lingenfelter, *Death Valley & the Amargosa: A Land of Illusion* (Berkeley: University of California, 1986), 42.

63 **a "bounteous stock of bread and beans":** William L. Manly, *Death Valley in '49: The Autobiography of a Pioneer*, reprint ed. (Santa Barbara, Calif.: The Narrative Press, 2001), 105.

64 **Most of the members of Manly's group:** Lingenfelter, *Death Valley & the Amargosa*, 34–35.

64 **a bloody liquid "resembling corruption":** Manly, *Death Valley in '49*, 106.

64 **asked him please to shut up:** Manly, *Death Valley in '49*, 99.

65 **"Creator's dumping place":** The account of this exchange comes from Manly, *Death Valley in '49*, 113.

65 **"enjoyed an extremely refreshing bath":** Cited in James E. Deacon and Cynthia Deacon Williams, "Ash Meadows and the Legacy of the Devils Hole Pupfish, in *Battle Against Extinction: Native Fish Management in the American West*, W. L. Minckley and James E. Deacon, eds. (Tucson: University of Arizona Press, 1991), 69.

65 **"not much more than an inch long":** Manly, *Death Valley in '49*, 107.

65 **a "beautiful enigma":** Christopher J. Norment, *Relicts of a Beautiful Sea: Survival, Extinction, and Conservation in a Desert World* (Chapel Hill: University of North Carolina, 2014), 110.

66 **fuzzy shots of two feet walking:** The surveillance video was posted with a story by Veronica Rocha, "3 Men Face Felony

Charges in Killing of Endangered Pupfish in Death Valley," *Los Angeles Times* (May 13, 2016), latimes.com/local/lanow/la-me-ln-pupfish-charges-20160513-snap-story.html.

68 **described him as potbellied and stern:** Paige Blankenbuehler, "How a Tiny Endangered Species Put a Man in Prison," *High Country News* (Apr. 15, 2019).

72 **weighed in at about a hundred grams:** This calculation is based on figures from Norment, *Relicts of a Beautiful Sea*, 120.

72 **"suitable for either ball or shot":** Manly, *Death Valley in '49*, 13.

72 **"the finest kind of food, fit for an epicure":** Manly, *Death Valley in '49*, 64.

72 **"Is it not a maimed and imperfect nature":** Henry David Thoreau, *Thoreau's Journals, Vol. 20* (entry from March 23, 1856), transcript available at: http://thoreau.library.ucsb.edu/writings_journals20.html.

73 **took place in 1882:** Joel Greenberg, *A Feathered River Across the Sky: The Passenger Pigeon's Flight to Extinction* (New York: Bloomsbury, 2014), 152–155.

73 **"It would have been as easy to count":** William T. Hornaday, *The Extermination of the American Bison with a Sketch of Its Discovery and Life History* (Washington, D.C.: Government Printing Office, 1889), 387.

73 **"hardly a bone will remain above ground":** Hornaday, *The Extermination of the American Bison*, 525.

73 **"For one species to mourn the death of another":** Aldo Leopold, *A Sand County Almanac*, reprint ed. (New York: Ballantine, 1970), 117.

74 **Extinction rates are now hundreds:** Anthony D. Barnosky et al., "Has the Earth's Sixth Mass Extinction Already Arrived?" *Nature*, 471 (2011) 51–57.

74 **a list of "common birds in steep decline":** The list, compiled by the U.S. North American Bird Conservation Initiative, is available at: allaboutbirds.org/news/state-of-the-birds-2014-common-birds-in-steep-decline-list/.

74 **Even among insects:** Caspar A. Hallmann et al., "More than 75

Percent Decline over 27 Years in Total Flying Insect Biomass in Protected Areas," *PLoS ONE*, 12 (2017), journals.plos.org/plosone/article?id=10.1371/journal.pone.0185809.

77 **The tests left behind a more or less permanent marker:** C. N. Waters et al., "Global Boundary Stratotype Section and Point (GSSP) for the Anthropocene Series: Where and How to Look for Potential Candidates," *Earth-Science Reviews*, 178 (2018), 379–429.

77 **"peculiar race of desert fish":** Proclamation 2961, 17 Fed. Reg. 691 (Jan. 23, 1952).

77 **That spring, the Department of Defense:** For a full list of nuclear tests by date, see U.S. Department of Energy, National Nuclear Safety Administration Nevada Field Office, *United States Nuclear Tests: July 1945 through September 1992* (Alexandria, Va.: U.S. Department of Commerce, 2015), nnss.gov/docs/docs_LibraryPublications/DOE_NV-209_Rev16.pdf.

77 **His plan was to construct from scratch:** This plan is described in Kevin C. Brown, *Recovering the Devils Hole Pupfish: An Environmental History* (National Park Service, 2017), 315. An electronic copy of the history was generously provided by the author.

78 **By the end of 1970:** Brown, *Recovering the Devils Hole Pupfish*, 142.

78 **the National Park Service rigged up a bank:** Brown, *Recovering the Devils Hole Pupfish*, 145.

78 **Some went to Saline Valley:** Brown, *Recovering the Devils Hole Pupfish*, 139.

79 **Then rival stickers appeared:** Brown, *Recovering the Devils Hole Pupfish*, 303.

79 **"Water, water, water":** Edward Abbey, *Desert Solitaire: A Season in the Wilderness*, reprint ed. (New York: Touchstone, 1990), 126.

81 **"All living things on earth are kindred":** Abbey, *Desert Solitaire*, 21.

81 **"To watch a small school of pupfish":** Norment, *Relicts of a Beautiful Sea*, 3–4.

82 **dubbed "misanthropic synanthropes":** Stanley D. Gehrt,

Justin L. Brown, and Chris Anchor, "Is the Urban Coyote a Misanthropic Synanthrope: The Case from Chicago," *Cities and the Environment*, 4 (2011), digitalcommons.lmu.edu/cate/vol4/iss1/3/.

83 **currently listed as "possibly extinct":** For the latest on the IUCN's list of "possibly extinct" animals, see: iucnredlist.org/statistics.

84 **The term of art for such creatures is "conservation-reliant":** J. Michael Scott et al., "Recovery of Imperiled Species under the Endangered Species Act: The Need for a New Approach, *Frontiers in Ecology and the Environment*, 3 (2005), 383–389.

84 **"Old deeds for old people":** Henry David Thoreau, *Walden*, reprint ed. (Oxford: Oxford University, 1997), 10.

85 **it is the "destiny of every considerable stream":** Mary Austin, *The Land of Little Rain*, reprint ed. (Mineola, N.Y.: Dover, 2015), 61.

85 **Among those creatures that lasted long enough:** Robert R. Miller, James D. Williams, and Jack E. Williams, "Extinctions of North American Fishes During the Past Century," *Fisheries*, 14 (1989), 22–38.

86 **"I distinctly remember being scared to death":** Edwin Philip Pister, "Species in a Bucket," *Natural History* (January 1993), 18.

87 **He managed to save thirty-two of them:** C. Moon Reed, "Only You Can Save the Pahrump Poolfish," *Las Vegas Weekly* (March 9, 2017), lasvegasweekly.com/news/2017/mar/09/pahrump-poolfish-lake-harriet-spring-mountain/.

88 **"Men make their own biosphere":** J. R. McNeill, *Something New Under the Sun: An Environmental History of the Twentieth-Century World* (New York: Norton, 2000), 194.

2

91 **something like half of the Caribbean's coral cover disappeared:** Richard B. Aronson and William F. Precht, "White-

Band Disease and the Changing Face of Caribbean Coral Reefs," *Hydrobiologia*, 460 (2001), 25–38.

91 **In 1998, a so-called global bleaching event:** Alexandra Witze, "Corals Worldwide Hit by Bleaching," *Nature* (Oct. 8, 2015), nature.com/news/corals-worldwide-hit-by-bleaching-1.18527.

91 **"stop growing and begin dissolving":** Jacob Silverman et al., "Coral Reefs May Start Dissolving When Atmospheric CO_2 Doubles," *Geophysical Research Letters*, 36 (2009), agupubs.online library.wiley.com/doi/full/10.1029/2008GL036282.

92 **"rapidly eroding rubble banks":** O. Hoegh-Guldberg et al., "Coral Reefs Under Rapid Climate Change and Ocean Acidification," *Science*, 318 (2007), 1737–1742.

95 **"curious rings of coral land":** Charles Darwin, *The Voyage of the Beagle* (New York: P. F. Collier, 1909), 406.

95 **"raised by myriads of tiny architects":** Darwin, *Charles Darwin's Beagle Diary*, Richard Darwin Keynes, ed. (Cambridge: Cambridge University, 1988), 418.

96 **"thirty-five folio pages of crabbed, elliptical scrawl":** Janet Browne, *Charles Darwin: Voyaging* (New York: Knopf, 1995), 437.

96 **"We see nothing of these slow changes":** Darwin, *On the Origin of Species: A Facsimile of the First Edition* (Cambridge, Mass.: Harvard University, 1964), 84.

96 **"Beneath this laurel's friendly pitying shade":** From an "Epitaph for a Favourite Tumbler Who Died Aged Twelve," signed Columba, full poem available at: darwinspigeons.com/#/victorian-pigeon-poems/4535732923.

97 **"retch awfully":** Darwin wrote this in a letter to his friend Thomas Eyton, cited in Browne, *Charles Darwin*, 525.

97 **"I have kept every breed":** Darwin, *On the Origin of Species*, 20–21.

97 **"If feeble man can do [so] much":** Darwin, *On the Origin of Species*, 109.

97 **after *The End of Nature*:** Bill McKibben, *The End of Nature* (New York: Random House, 1989).

98 **more than ninety percent of the Great Barrier Reef:** This

figure comes from Neal Cantin, a research scientist I interviewed at the SeaSim (Nov. 15, 2019).

98 **half its corals had perished:** Robinson Meyer, "Since 2016, Half of All Coral in the Great Barrier Reef Has Died," *The Atlantic* (Apr. 18, 2018), theatlantic.com/science/archive/2018/04/since-2016-half-the-coral-in-the-great-barrier-reef-has-perished/558302/.

98 **a "catastrophic" collapse:** Terry P. Hughes et al., "Global Warming Transforms Coral Reef Assemblages," *Nature*, 556 (2018), 492–496.

105 **a healthy patch of reef:** Mark D. Spalding, Corinna Ravilious, and Edmund P. Green, *World Atlas of Coral Reefs* (Berkeley: University of California, 2001), 27.

105 **Researchers once picked apart:** Spalding et al., *World Atlas of Coral Reefs*, 27.

105 **Using genetic-sequencing techniques:** Laetitia Plaisance et al., "The Diversity of Coral Reefs: What Are We Missing?" *PLoS ONE*, 6 (2011), journals.plos.org/plosone/article?id=10.1371/journal.pone.0025026.

105 **between one and nine million species:** Nancy Knowlton, "The Future of Coral Reefs," *Proceedings of the National Academy of Sciences*, 98 (2001), 5419–5425.

106 **"In the coral city there is no waste":** Richard C. Murphy, *Coral Reefs: Cities under the Sea* (Princeton, N.J.: The Darwin Press, 2002), 33.

106 **"It will be slimy":** Roger Bradbury, "A World Without Coral Reefs," *The New York Times* (July 13, 2012), A17.

106 **The authority said that the reef's long-term prospects:** Great Barrier Reef Marine Park Authority, *Great Barrier Reef Outlook Report 2019* (Townsville, Aus.: GBRMPA, 2019), vi. The full report is available at: http://elibrary.gbrmpa.gov.au/jspui/handle/11017/3474/.

106 **a gigantic new coal mine:** "Adani Gets Final Environmental Approval for Carmichael Mine," *Australian Broadcasting Corporation News* (last updated June 13, 2019), abc.net.au/news/2019

-06-13/adani-carmichael-coal-mine-approved-water-management-galilee/11203208.

107 **"The world's most insane energy project":** Jeff Goodell, "The World's Most Insane Energy Project Moves Ahead," *Rolling Stone* (June 14, 2019), rollingstone.com/politics/politics-news/adani-mine-australia-climate-change-848315/.

109 **"entangled bank, clothed with many plants of many kinds":** Darwin, *On the Origin of Species*, 489.

3

115 **calls himself a "genetic designer":** Josiah Zayner, "How to Genetically Engineer a Human in Your Garage—Part I," josiahzayner.com/2017/01/genetic-designer-part-i.html.

116 **"a way to rewrite the very molecules of life":** Jennifer A. Doudna and Samuel H. Sternberg, *A Crack in Creation: Gene Editing and the Unthinkable Power to Control Evolution* (Boston: Houghton Mifflin Harcourt, 2017), 119.

116 **ants that can't smell:** Waring Trible et al, "*orco* Mutagenesis Causes Loss of Antennal Lobe Glomeruli and Impaired Social Behavior in Ants," *Cell*, 170 (2017), 727–735.

116 **macaques that suffer from sleep disorders:** Peiyuan Qiu et al., "BMAL1 Knockout Macaque Monkeys Display Reduced Sleep and Psychiatric Disorders," *National Science Review*, 6 (2019), 87–100.

116 **Eadweard Muybridge's famous series of photographs:** Seth L. Shipman et al., "CRISPR-Cas Encoding of a Digital Movie into the Genomes of a Population of Living Bacteria," *Nature*, 547 (2017), 345–349.

117 **The Australian Animal Health Laboratory:** Several months after I visited, the Australian Animal Health Laboratory was renamed the Australian Centre for Disease Preparedness.

119 **"an enormous, warty bufonid":** U.S. Fish and Wildlife Service, "Cane Toad (*Rhinella marina*) Ecological Risk Screening Sum-

mary," web version (revised Apr. 5, 2018), fws.gov/fisheries/ans/erss/highrisk/ERSS-Rhinella-marina-final-April2018.pdf.

119 **"Large individuals sitting on roadways":** L. A. Somma, "Rhinella marina (Linnaeus, 1758)," U.S. Geological Survey, *Nonindigenous Aquatic Species Database* (revised Apr. 11, 2019), nas.er.usgs.gov/queries/FactSheet.aspx?SpeciesID=48.

119 **A toad named Bette Davis:** Rick Shine, *Cane Toad Wars* (Oakland: University of California, 2018), 7.

120 **In the mid-1800s, they were imported to the Caribbean:** Byron S. Wilson et al., "Cane Toads a Threat to West Indian Wildlife: Mortality of Jamaican Boas Attributable to Toad Ingestion," *Biological Invasions,* 13 (2011), link.springer.com/article/10.1007/s10530-010-9787-7.

120 **they'd produced more than 1.5 million eggs:** Shine, *Cane Toad Wars,* 21.

120 **toads on the front lines had significantly longer legs:** Benjamin L. Phillips et al., "Invasion and the Evolution of Speed in Toads," *Nature*, 439 (2006), 803.

121 **"It has invaded the Territory":** Karen Michelmore, "Super Toad," *Northern Territory News* (Feb. 16, 2006), 1.

122 **The list of species whose numbers have crashed:** Shine, *Cane Toad Wars,* 4. See also: "The Biological Effects, Including Lethal Toxic Ingestion, Caused by Cane Toads (Bufo marinus): Advice to the Minister for the Environment and Heritage from the Threatened Species Scientific Committee (TSSC) on Amendments to the List of Key Threatening Processes under the Environment Protection and Biodiversity Conservation Act 1999 (EPBC Act)" (Apr. 12, 2005), environment.gov.au/biodiversity/threatened/key-threatening-processes/biological-effects-cane-toads.

123 **the Australian government offer a bounty:** House of Representatives Standing Committee on the Environment and Energy, *Cane Toads on the March: Inquiry into Controlling the Spread of Cane Toads* (Canberra: Commonwealth of Australia, 2019), 32.

126 **amplifies the poison's potency a hundredfold:** Robert Capon,

"Inquiry into Controlling the Spread of Cane Toads, Submission 8" (Feb. 2019). Available for download at: aph.gov.au/Parliamentary_Business/Committees/House/Environment_and_Energy/Canetoads/Submissions.

127 **Feed them toad "sausages":** Naomi Indigo et al., "Not Such Silly Sausages: Evidence Suggests Northern Quolls Exhibit Aversion to Toads after Training with Toad Sausages," *Austral Ecology*, 43 (2018), 592–601.

128 **Some interfere with the replication of a rival gene:** Austin Burt and Robert Trivers, *Genes in Conflict: The Biology of Selfish Genetic Elements* (Cambridge, Mass.: Belknap, 2006), 4–5.

128 **passed on more than ninety percent of the time:** Burt and Trivers, *Genes in Conflict*, 3.

128 **including mosquitoes, flour beetles, and lemmings:** Burt and Trivers, *Genes in Conflict*, 13–14.

129 **to create a synthetic gene drive in yeast:** James E. DiCarlo et al., "Safeguarding CRISPR-Cas9 Gene Drives in Yeast," *Nature Biotechnology*, 33 (2015), 1250–1255.

129 **to create a synthetic gene drive in fruit flies:** Valentino M. Gantz and Ethan Bier, "The Mutagenic Chain Reaction: A Method for Converting Heterozygous to Homozygous Mutations," *Science*, 348 (2015), 442–444.

131 **until yellow ruled:** Doudna and Sternberg estimate that had the gene-drive fruit flies escaped, they could have spread the gene for yellow coloring to between a fifth and a half of all fruit flies worldwide. *A Crack in Creation*, 151.

132 **"There is hope":** GBIRd website, geneticbiocontrol.org.

133 **down to zero within a few years:** Thomas A. A. Prowse, et al., "Dodging Silver Bullets: Good CRISPR Gene-Drive Design Is Critical for Eradicating Exotic Vertebrates," *Proceedings of the Royal Society B*, 284 (2017), royalsocietypublishing.org/doi/10.1098/rspb.2017.0799.

134 **claimed at least a thousand species of island birds:** Richard P. Duncan, Alison G. Boyer, and Tim M. Blackburn, "Magnitude and Variation of Prehistoric Bird Extinctions in the Pacific,"

Proceedings of the National Academy of Sciences, 110 (2013), 6436–6441.

134 **Despite intensive efforts to save them:** Elizabeth A. Bell, Brian D. Bell, and Don V. Merton, "The Legacy of Big South Cape: Rat Irruption to Rat Eradication," *New Zealand Journal of Ecology*, 40 (2016), 212–218.

134 **"Only humans are as adaptable":** Lee M. Silver, *Mouse Genetics: Concepts and Applications* (Oxford: Oxford University, 1995), adapted for the Web by Mouse Genome Informatics, The Jackson Laboratory (revised Jan. 2008), http://informatics.jax.org/silver/.

134 **"like working in an ornithological trauma center":** Alex Bond, "Mice Wreak Havoc for South Atlantic Seabirds," *British Ornithologists' Union*, bou.org.uk/blog-bond-gough-island-mice-seabirds/.

135 **compared to Kurt Vonnegut's *ice-nine*:** Rowan Jacobsen, "Deleting a Species," *Pacific Standard* (June–July 2018, updated Sept. 7, 2018), psmag.com/magazine/deleting-a-species-genetically-engineering-an-extinction.

135 **with names like "killer-rescue":** Jaye Sudweeks et al., "Locally Fixed Alleles: A Method to Localize Gene Drive to Island Populations," *Scientific Reports*, 9 (2019), doi.org/10.1038/s41598-019-51994-0.

136 **featuring a so-called CATCHA sequence:** Bing Wu, Liqun Luo, and Xiaojing J. Gao, "Cas9-Triggered Chain Ablation of *Cas9* as Gene Drive Brake," *Nature Biotechnology*, 34 (2016), 137–138.

138 **"new techniques of genetic rescue":** Revive & Restore website, reviverestore.org/projects/.

139 **"Do you know how he did it?":** Dr. Seuss, *The Cat in the Hat Comes Back* (New York: Beginner Books, 1958), 16.

139 **"an extinction avalanche":** Edward O. Wilson, *The Future of Life* (New York: Vintage, 2002), 53.

139 **"We are not as gods":** Wilson, *Half-Earth: Our Planet's Fight for Life* (New York: Liveright, 2016), 51.

139 **"We are as gods, but we have failed to get good at it":** Paul

Kingsnorth, "Life Versus the Machine," *Orion* (Winter 2018), 28–33.

Up in the Air

1

147 **"The start of the switchover":** William F. Ruddiman, *Plows, Plagues, and Petroleum: How Humans Took Control of Climate* (Princeton, N.J.: Princeton University, 2005), 4.

147 **humans emitted some fifteen million tons of CO_2:** Historical emissions data come from Hannah Ritchie and Max Roser, "CO_2 and Greenhouse Gas Emissions," *Our World in Data* (last revised Aug. 2020), ourworldindata.org/CO2-and-other-greenhouse-gas-emissions.

148 **Droughts are growing deeper:** Benjamin Cook, "Climate Change Is Already Making Droughts Worse," *CarbonBrief* (May 14, 2018), carbonbrief.org/guest-post-climate-change-is-already-making-droughts-worse.

148 **storms fiercer:** Kieran T. Bhatia et al., "Recent Increases in Tropical Cyclone Intensification Rates," *Nature Communications*, 10 (2019), doi.org/10.1038/s41467-019-08471-z.

148 **Wildfire season is getting longer:** W. Matt Jolly et al., "Climate-Induced Variations in Global Wildfire Danger from 1979 to 2013," *Nature Communications*, 6 (2015), doi.org/10.1038/ncomms8537.

148 **melt off of Antarctica has increased threefold:** A. Shepherd et al., "Mass Balance of the Antarctic Ice Sheet from 1992 to 2017," *Nature*, 558 (2018), 219–222.

148 **most atolls will, in another few decades:** Curt D. Storlazzi et al., "Most Atolls Will Be Uninhabitable by the Mid-21st Century Because of Sea-Level Rise Exacerbating Wave-Driven Flooding," *Science Advances*, 25 (2018), advances.sciencemag.org/content/4/4/eaap9741.

148 **"holding the increase in the global average temperature":** The full text of the Paris Agreement in English is available at: unfccc.int/files/essential_background/convention/application/pdf/english_paris_agreement.pdf.

148 **To stave off 1.5°C:** There are many ways to calculate how much CO_2 can still be emitted if the world is to stay below 1.5° or 2°C; I am using the Mercator Research Institute on Global Commons and Climate Change's "remaining carbon budget" figures, available at: mcc-berlin.net/en/research/CO2-budget.html.

150 **"smaller than many deserts":** K. S. Lackner and C. H. Wendt, "Exponential Growth of Large Self-Reproducing Machine Systems," *Mathematical and Computer Modelling,* 21 (1995), 55–81.

151 **as "adventure capital":** Wallace S. Broecker and Robert Kunzig, *Fixing Climate: What Past Climate Changes Reveal About the Current Threat—and How to Counter It* (New York: Hill and Wang, 2008), 205.

152 **"Rewarding people for going to the bathroom less":** Klaus S. Lackner and Christophe Jospe, "Climate Change Is a Waste Management Problem," *Issues in Science and Technology,* 33 (2017), issues.org/climate-change-is-a-waste-management-problem/.

153 **"Such a moral stance":** Lackner and Jospe, "Climate Change Is a Waste Management Problem."

153 **global CO_2 emissions were down:** Chris Mooney, Brady Dennis, and John Muyskens, "Global Emissions Plunged an Unprecedented 17 Percent during the Coronavirus Pandemic," *The Washington Post* (May 19, 2020), washingtonpost.com/climate-environment/2020/05/19/greenhouse-emissions-coronavirus/?arc404=true.

153 **How long, exactly, is a complicated question:** Individual carbon molecules are constantly cycling between atmosphere and oceans and between both of these and the world's vegetation. However, CO_2 levels in the atmosphere are governed by much slower processes. For a fuller discussion, see Doug Mackie, "CO_2 Emissions Change Our Atmosphere for Centuries," *Skeptical Sci-*

ence (last updated July 5, 2015), skepticalscience.com/argument.php?p=1&t=77&&a=80.

154 **the United States is responsible:** All figures on aggregate emissions are taken from Hannah Ritchie, "Who Has Contributed Most to Global CO_2 Emissions?" *Our World in Data* (Oct. 1, 2019), ourworldindata.org/contributed-most-global-CO2.

155 **a hundred and one involved negative emissions:** Sabine Fuss et al., "Betting on Negative Emissions," *Nature Climate Change*, 4 (2014), 850–852.

155 ***All*** **of the scenarios consistent with that goal:** J. Rogelj et al., "Mitigation Pathways Compatible with 1.5°C in the Context of Sustainable Development," in *Global Warming of 1.5°C: An IPCC Special Report*, V. Masson-Delmotte et al., eds., Intergovernmental Panel on Climate Change (Oct. 8, 2018), ipcc.ch/site/assets/uploads/sites/2/2019/02/SR15_Chapter2_Low_Res.pdf.

158 **I used up my allotment:** Calculating the emissions from air travel is complicated, and different groups offer different estimates for the same trip. I am relying on the flight carbon calculator at myclimate.org.

159 **A recent study by Swiss researchers:** Jean-Francois Bastin et al., "The Global Tree Restoration Potential," *Science*, 364 (2019), 76–79.

159 **Other researchers argued:** Katarina Zimmer, "Researchers Find Flaws in High-Profile Study on Trees and Climate," *The Scientist* (Oct. 17, 2019), the-scientist.com/news-opinion/researchers-find-flaws-in-high-profile-study-on-trees-and-climate--66587. DOI: 10.1126/science.aay7976.

159 **was "still substantial":** Joseph W. Veldman et al., "Comment on 'The Global Tree Restoration Potential,'" *Science*, 366 (2019), science.sciencemag.org/content/366/6463/eaay7976.

159 **One entails cutting down mature trees:** Ning Zeng, "Carbon Sequestration Via Wood Burial," *Carbon Balance and Management*, 3 (2008), doi.org/10.1186/1750-0680-3-1.

159 **Another scheme involves collecting crop residues:** Stuart E. Strand and Gregory Benford, "Ocean Sequestration of Crop Resi-

due Carbon: Recycling Fossil Fuel Carbon Back to Deep Sediments," *Environmental Science and Technology*, 43 (2009), 1000–1007.

160 **"Assuming it takes a crew of ten people":** Zeng, "Carbon Sequestration Via Wood Burial."

160 **a recent study by a team of German scientists:** Jessica Strefler et al., "Potential and Costs of Carbon Dioxide Removal by Enhanced Weathering of Rocks," *Environmental Research Letters* (March 5, 2018), dx.doi.org/10.1088/1748-9326/aaa9c4.

161 **"two steps backward in justice":** Olúfẹ́mi O. Táíwò, "Climate Colonialism and Large-Scale Land Acquisitions," *C2G* (Sept. 26, 2019), c2g2.net/climate-colonialism-and-large-scale-land-acquisitions/.

2

166 **reached a height of twenty-five miles:** Clive Oppenheimer, *Eruptions that Shook the World* (New York: Cambridge University, 2011), 299.

166 **Ten thousand people were killed:** Oppenheimer, *Eruptions that Shook the World*, 310.

166 **"a body of liquid fire":** The account of the Rajah of Sanggar is cited in Oppenheimer, *Eruptions that Shook the World*, 299.

166 **"It was impossible to see your hand":** This account, from the captain of a ship owned by the East India Company, is cited in Gillen D'Arcy Wood, *Tambora: The Eruption that Changed the World* (Princeton, N.J.: Princeton University, 2014), 21.

166 **more than a hundred million tons of gas:** South Dakota State University, "Undocumented Volcano Contributed to Extremely Cold Decade from 1810–1819," *ScienceDaily* (Dec. 7, 2009), sciencedaily.com/releases/2009/12/091205105844.htm.

167 **"ruined figures, scarcely resembling men":** Cited in Oppenheimer, *Eruptions that Shook the World*, 314.

167 **marching under the banner BREAD OR BLOOD:** William K. Klingaman and Nicholas P. Klingaman, *The Year Without Sum-*

mer: 1816 and the Volcano That Darkened the World and Changed History (New York: St. Martin's, 2013), 46.

167 **some estimates put the figure in the millions:** Wood, *Tambora*, 233.

168 **"The very face of nature":** Cited in Klingaman and Klingaman, *The Year Without Summer*, 64.

168 **On July 8, there was frost:** Klingaman and Klingaman, *The Year Without Summer*, 104.

168 **Chester Dewey, a professor at Williams College:** Cited in Oppenheimer, *Eruptions that Shook the World*, 312.

169 **"dangerous beyond belief":** James Rodger Fleming, *Fixing the Sky: The Checkered History of Weather and Climate Control* (New York: Columbia University, 2010), 2.

169 **"a broad highway to hell":** This assessment comes from Tim Flannery, cited in Mark White, "The Crazy Climate Technofix," *SBS* (May 27, 2016), sbs.com.au/topics/science/earth/feature/geoengineering-the-crazy-climate-technofix.

169 **"unimaginably drastic":** Holly Jean Buck, *After Geoengineering: Climate Tragedy, Repair, and Restoration* (London: Verso, 2019), 3.

169 **and also as "inevitable":** Dave Levitan, "Geoengineering Is Inevitable," *Gizmodo* (Oct. 9, 2018), earther.gizmodo.com/geoengineering-is-inevitable-1829623031.

171 **a brief downturn in global temperatures:** "Global Effects of Mount Pinatubo," *NASA Earth Observatory* (June 15, 2001), earthobservatory.nasa.gov/images/1510/global-effects-of-mount-pinatubo.

171 **In the tropics, ozone levels in the lower stratosphere:** William B. Grant et al., "Aerosol-Associated Changes in Tropical Stratospheric Ozone Following the Eruption of Mount Pinatubo," *Journal of Geophysical Research*, 99 (1994), 8197–8211.

172 **"Man is unwittingly conducting":** President's Science Advisory Committee, *Restoring the Quality of Our Environment: Report of the Environmental Pollution Panel* (Washington, D.C.: The White House, 1965), 126.

173 **"rise about four feet every ten years":** *Restoring the Quality of Our Environment*,123.

173 **"Rough estimates indicate":** *Restoring the Quality of Our Environment*, 127.

173 **sending aircraft to seed the clouds:** H. E. Willoughby et al., "Project STORMFURY: A Scientific Chronicle 1962–1983," *Bulletin of the American Meteorological Society*, 66 (1985), 505–514.

173 **An astonishing twenty-six hundred seeding sorties:** Fleming, *Fixing the Sky*, 180.

174 **Other climate-modification plans pursued:** National Research Council, *Weather & Climate Modification: Problems and Progress* (Washington, D.C.: The National Academies Press, 1973), 9.

175 **"What mankind needs is a war against cold":** Cited in Fleming, *Fixing the Sky*, 202.

175 **Gorodsky believed this arrangement:** Nikolai Rusin and Liya Flit, *Man Versus Climate*, Dorian Rottenberg, trans. (Moscow: Peace Publishers, 1962), 61–63.

175 **"New projects for transforming nature":** Rusin and Flit, *Man Versus Climate*, 174.

175 **Public concern about the environment:** David W. Keith, "Geoengineering the Climate: History and Prospect," *Annual Review of Energy and the Environment*, 25 (2000), 245–284.

176 **"rockets and different types of missiles":** Mikhail Budyko, *Climatic Changes*, American Geophysical Union, trans. (Baltimore: Waverly, 1977), 241.

176 **"climate modification will become necessary":** Budyko, *Climatic Changes*, 236.

176 **"foremost proponent of geoengineering":** Joe Nocera, "Chemo for the Planet," *The New York Times* (May 19, 2015), A25.

176 **"I'm a proponent of reality":** David Keith, Letter to the Editor, *The New York Times* (May 27, 2015), A22.

177 **he describes himself as a "tinkerer":** David Keith, *A Case for Climate Engineering* (Cambridge, Mass.: MIT, 2013), xiii.

179 **development costs would run to about $2.5 billion:** Wake Smith and Gernot Wagner, "Stratospheric Aerosol Injection Tactics and Costs in the First 15 Years of Deployment," *Environmental Research Letters*, 13 (2018), doi.org/10.1088/1748-9326/aae98d.

180 **three hundred times that amount every year:** It's been estimated that global fossil-fuel subsidies totaled $5.2 trillion in 2017; see: David Coady et al., "Global Fossil Fuel Subsidies Remain Large: An Update Based on Country-Level Estimates," *IMF* (May 2, 2019), imf.org/en/Publications/WP/Issues/2019/05/02/Global-Fossil-Fuel-Subsidies-Remain-Large-An-Update-Based-on-Country-Level-Estimates-46509.

180 **"Dozens of countries would have both":** Smith and Wagner, "Stratospheric Aerosol Injection Tactics and Costs."

181 **the number of flights would ramp up accordingly:** Smith and Wagner, "Stratospheric Aerosol Injection Tactics and Costs."

181 **determined it would change the appearance of the sky:** Ben Kravitz, Douglas G. MacMartin, and Ken Caldeira, "Geoengineering: Whiter Skies?" *Geophysical Research Letters*, 39 (2012), doi.org/10.1029/2012GL051652.

181 **the latest version has more than two dozen entries:** Alan Robock, "Benefits and Risks of Stratospheric Solar Radiation Management for Climate Intervention (Geoengineering)," *The Bridge* (Spring 2020), 59–67.

184 **"Ironically, such engineering efforts":** Dan Schrag, "Geobiology of the Anthropocene," in *Fundamentals of Geobiology*, Andrew H. Knoll, Donald E. Canfield, and Kurt O. Konhauser, eds. (Oxford: Blackwell Publishing, 2012), 434.

3

187 **"Iceworm thus couples mobility":** Cited in Erik D. Weiss, "Cold War Under the Ice: The Army's Bid for a Long-Range

Nuclear Role, 1959–1963," *Journal of Cold War Studies*, 3 (2001), 31–58.

188 **"Camp Century is a symbol of man's unceasing struggle":** *The Story of Camp Century: The City Under Ice* (U.S. Army film 1963, digitized version 2012).

188 **two Boy Scouts—one American, one Danish:** Ronald E. Doel, Kristine C. Harper, and Matthias Heymann, "Exploring Greenland's Secrets: Science, Technology, Diplomacy, and Cold War Planning in Global Contexts," in *Exploring Greenland: Cold War Science and Technology on Ice*, Ronald E. Doel, Kristine C. Harper, and Matthias Heymann, eds. (New York: Palgrave, 2016), 16.

188 **Almost at once, the tunnels started to contract:** Kristian H. Nielsen, Henry Nielsen, and Janet Martin-Nielsen, "City Under the Ice: The Closed World of Camp Century in Cold War Culture," *Science as Culture*, 23 (2014), 443–464.

188 **the annual general meeting of all the devils of hell:** Willi Dansgaard, *Frozen Annals: Greenland Ice Cap Research* (Odder, Denmark: Narayana Press, 2004), 49.

189 **more than a thousand in all:** Jon Gertner, *The Ice at the End of the World: An Epic Journey Into Greenland's Buried Past and Our Perilous Future* (New York: Random House, 2019), 202.

194 **didn't seem to realize what a "gold mine" of data:** Dansgaard, *Frozen Annals*, 55.

194 **Dansgaard's reading of the Camp Century core:** W. Dansgaard et al., "One Thousand Centuries of Climatic Record from Camp Century on the Greenland Ice Sheet," *Science*, 166 (1969), 377–380.

195 **"a three-year-old who has just discovered a light switch":** Richard B. Alley, *The Two-Mile Time Machine: Ice Cores, Abrupt Climate Change, and Our Future* (Princeton: Princeton University, 2000), 120.

195 **it was a doozy:** Alley, *The Two-Mile Time Machine*, 114.

198 **temperatures on the ice sheet:** These figures come from Konrad Steffen, who tragically died in an accident on the ice sheet just

as this book was going to press. They are cited in: Gertner, "In Greenland's Melting Ice, A Warning on Hard Climate Choices," *e360* (June 27, 2019), e360.yale.edu/features/in-greenlands-melting-ice-a-warning-on-hard-climate-choices.

198 **ice loss from Greenland has increased sevenfold:** A. Shepherd et al., "Mass Balance of the Greenland Ice Sheet from 1992 to 2018," *Nature*, 579 (2020), 233–239.

198 **during an exceptionally warm couple of days:** Marco Tedesco and Xavier Fettweis, "Unprecedented Atmospheric Conditions (1948–2019) Drive the 2019 Exceptional Melting Season over the Greenland Ice Sheet," *The Cryosphere*, 14 (2020), 1209–1223.

198 **Greenland shed almost six hundred billion tons of ice:** Ingo Sasgen et al., "Return to Rapid Ice Loss in Greenland and Record Loss in 2019 Detected by GRACE-FO Satellites," *Communications Earth & Environment*, 1 (2020), doi.org/10.1038/s43247-020-0010-1.

198 **"The current Arctic is experiencing rates":** Eystein Jansen et al., "Past Perspectives on the Present Era of Abrupt Arctic Climate Change," *Nature Climate Change*, 10 (2020), 714–721.

199 **An early cost estimate for the project:** Peter Dockrill, "U.S. Army Weighs Up Proposal For Gigantic Sea Wall to Defend N.Y. from Future Floods," *ScienceAlert* (Jan. 20, 2020), sciencealert.com/storm-brewing-over-giant-6-mile-sea-wall-to-defend-new-york-from-future-floods.

199 **"We understand the hesitancy":** John C. Moore et al., "Geoengineer Polar Glaciers to Slow Sea-Level Rise," *Nature*, 555 (2018), 303–305.

200 **"We live in a world":** Andy Parker is quoted in Brian Kahn, "No, We Shouldn't Just Block Out the Sun," *Gizmodo* (Apr. 24, 2020), earther.gizmodo.com/no-we-shouldnt-just-block-out-the-sun-1843043812. I have undeleted the expletive.

Afterword

204 **"Last impressions may be lasting impressions":** Donald A. Redelmeier et al., "Memories of Colonoscopy: A Randomized Trial," *Pain*, 104 (2003), 187–194.

205 **had the proper protocols been followed:** David Cyranoski, "What China's Coronavirus Response Can Teach the Rest of the World," *Nature* (March 17, 2020), nature.com/articles/d41586-020-00741-x.

205 **"We live at high densities in many cities":** David Quammen, *Spillover: Animal Infections and the Next Human Pandemic* (New York: Norton, 2012), 515–516.

206 **"like trying to plan the construction of a stepping-stone pathway to the Moon":** Michael Osterholm, quoted by Nicky Phillips, "The Coronavirus Is Here to Stay—Here's What That Means," *Nature* (Feb. 16, 2021), nature.com/articles/d41586-021-00396-2.

Credits

Page 10 MGMT. design
11 MGMT. design
20 MGMT. design
22 © Ryan Hagerty, U.S. Fish and Wildlife Service
34 MGMT. design
39 © Drew Angerer/Getty Images
45 The Historic New Orleans Collection, 1974.25.11.2
57 © Danita Delimont/Alamy Stock Photo
69 National Park Service Photo by Brett Seymour/Submerged Resources Center
71 MGMT. design, adapted from Alan C. Riggs and James E. Deacon, "Connectivity in Desert Aquatic Ecosystems: The Devils Hole Story."
78, 79 Photos by Phil Pister, California Department of Fish and Wildlife and Desert Fishes Council, Bishop, CA.
96 Originally published in Charles Darwin, *Animals and Plants Under Domestication,* vol. 1.

99 MGMT. design
102 Photo: © Wilfredo Licuanan, courtesy of Corals of the World, coralsoftheworld.org.
111 © James Craggs, Horniman Museum and Gardens
121 MGMT. design
123 Photo: Arthur Mostead Photography, AMPhotography.com.au
125 MGMT. design
130 MGMT. design
146 Courtesy of U.S. Department of Energy/Pacific Northwest National Laboratory
154 MGMT. design, adapted from Zeke Hausfather, based on data from *Global Warming of 1.5°C: An IPCC Special Report.*
156 MGMT. design, adapted from *Global Warming of 1.5°C: An IPCC Special Report,* figure 2.5.
162 MGMT. design
166 © Iwan Setiyawan/AP Photo/KOMPAS Images
170 MGMT. design
174 Courtesy of soviet-art.ru.
178 MGMT. design, adapted from David Keith
190 Photo by Pictorial Parade/Archive Photos/Getty Images
191 Photo by US Army/Pictorial Parade/Archive Photos/Getty Images
192 MGMT. design
196 MGMT. design, adapted from Kurt M. Cuffey and Gary D. Clow, "Temperature, Accumulation, and Ice Sheet Elevation in Central Greenland Through the Last Deglacial Transition," *Journal of Geophysical Research* 102 (1997).

Photo: © John Kleiner

ELIZABETH KOLBERT is a staff writer at *The New Yorker* and the author of *The Sixth Extinction*, which won the Pulitzer Prize for general nonfiction in 2015. She's a two-time National Magazine Award winner. Her work has also been honored with a National Academies Communication Award, the Blake-Dodd Prize from the American Academy of Arts and Letters, a Heinz Award, and a Guggenheim Fellowship. Kolbert is a visiting fellow at Williams College's Center for Environmental Studies. She lives in Williamstown, Massachusetts, with her husband and children.

About the Type

The text of this book was set in Janson, a typeface designed about 1690 by Nicholas Kis (1650–1702), a Hungarian living in Amsterdam, and for many years mistakenly attributed to the Dutch printer Anton Janson. In 1919, the matrices became the property of the Stempel Foundry in Frankfurt. It is an old-style book face of excellent clarity and sharpness. Janson serifs are concave and splayed; the contrast between thick and thin strokes is marked.

ALSO AVAILABLE FROM PULITZER PRIZE–WINNING AUTHOR ELIZABETH KOLBERT

The book that launched Elizabeth Kolbert's career as an environmental writer—"among the few irreplaceable volumes yet written about climate change."

—Bill McKibben, *The Boston Globe*